"I whole-heartedly recommend Catherine G. L[ ] of *Spiritual Emergency* as an enormously help[ ] path. Written from personal experience, it adds credibility and comp[ ]derstanding of the challenges that almost all those who undertake an inner journey of self-discovery and awakening encounter — and welcome practical help for dealing with and overcoming them."

> DR. JUDE CURRIVAN,
> author of *The 8th Chakra, The 13th Step* and *HOPE – Healing Our People &*
> *Earth*; and co-author with Dr. Ervin Laszlo of *CosMos. www.judecurrivan.com*

"This book is a remarkable achievement. The author charts the terrain of moving from breakdown to breakthrough, exploring how surviving a mental health crisis can lead to a new and authentic way of being and bringing a profound message of hope to all sufferers. The lesson for mental health professionals is clear: to recognise that mental health is not to be equated with social norms and conventions. Rather, we should do our utmost to help each individual feel validated as they journey through breakdown towards the possibility of re-visioning their world, respecting their unique gifts and affirming the spiritual values that inspire them to continue on the path of life."

> DR. ANDREW POWELL,
> *Founding Chair, Spirituality and Psychiatry Special Interest Group, Royal College*
> *of Psychiatrists UK.*

"A highly relevant book at this time of unprecedented spiritual awakening. A great Road Map to safely become who you really are — whilst understanding and avoiding the pitfalls."

> HAZEL COURTENEY,
> author of *Countdown to Coherence, Divine Intervention* and
> *The Evidence for the Sixth Sense.*

"Catherine G. Lucas has brought us tremendous insight from both personal experience and professional reflection that combine to make this book an excellent resource for anyone working in the area of Spiritual Care and Mental Health. I have already found it to be of great assistance as a spiritual care-giver, and I have no doubt that this delightfully accessible yet well grounded book will be of great significance and help to anyone who seeks guidance in the area of spiritual emergency."

> NICK JONES,
> *Mental Health Chaplain and Spiritual Care Co-ordinator –*
> *Black Country Partnership Foundation Trust.*

IN CASE OF
SPIRITUAL EMERGENCY

Moving Successfully Through
Your Awakening

Catherine G Lucas

FINDHORN PRESS

The right of Catherine G. Lucas to be identified as the author
of this work has been asserted by her in accordance
with the Copyright, Designs and Patents Act 1998.

Published in 2011 by Findhorn Press, Scotland

ISBN 978-1-84409-546-9

A CIP record for this title is available from the British Library.

Edited by Patricia Kot
Cover design by Richard Crookes
Interior design by Damian Keenan
Printed and bound in the European Union

1 2 3 4 5 6 7 8 9 17 16 15 14 13 12 11

Published by
Findhorn Press
117-121 High Street,
Forres IV36 1AB,
Scotland, UK
t +44 (0)1309 690582
f +44 (0)131 777 2711
e info@findhornpress.com
www.findhornpress.com

CONTENTS

Part II
Spiritual Emergency Through the Ages

Part III
Moving Successfully Through Spiritual Emergency: The Three Key Phases

For Swithin, my beloved
For Stewart, my blessing
And for all those going through spiritual emergency

Acknowledgments

Heart-felt thanks to Aaron's mother, Al, Annabel, Emily, Emma, Jennifer, Kate and Kimberley for sharing your personal stories so generously and so openly for the benefit of others; truly an act of service. Warm thanks also to members of the Stroud Spiritual Crisis Network group for your ongoing support and extremely helpful discussion of parts of the manuscript.

Thanks also to Andrew Powell for your invaluable feedback and suggestions when I was grappling with the two medical and scientific chapters. And to Hazel Courteney for your encouragement and for alerting me to the two news stories covered in the book.

I would like to thank Vidyamala too, my awesome and inspiring Mindfulness teacher. The chapter on Mindfulness would not have been possible without you. And I am grateful to the whole Findhorn Press team for having the vision to believe in the book and especially to Sabine for skilfully guiding me through the whole publication process.

I want to end by thanking the Universe, for also guiding me through the whole process. And for sending me a life-partner who turned out to be not only a fellow writer, but also a professional, trusty sub-editor. And one who has the patience of a saint. You have put up with far more than could reasonably be expected of any fiancé! Thank you all.

ANNABEL facilitates groups on the theme of creativity and the cycle of the year, as well as a dream group. She is a massage therapist and a founding member of the Spiritual Crisis Network.

KIMBERLEY is an Energy Intuitive, Reiki Master Healer, writer and intuitive artist. She is the founder of The e-Wakening Academy and One Minute for Peace Quantum Peace Project.

JENNIFER is a licensed psychologist who has studied, practiced, and taught psychology since 1969. She works as a school psychologist in early intervention with three- and four-year-olds, leads art retreats and facilitates a program for sharing creativity. She paints as a form of prayer. Jennifer is the author of *Dancing with God through the Storm: Mysticism and Mental Illness* (Way Opens Press 2002). A Quaker, she lives in Pennsylvania, in a community.

KATE works in digital media.

EMMA runs an ethical film making business that helps to promote healing and the raising of consciousness. She is also a Shiatsu Practitioner.

AL now enjoys working as a volunteer after a long career in public service.

EMILY is a Quaker and lives in the south of England.

Looking after yourself

Many of you reading this book will not currently be in the midst of spiritual emergency, but if you are, please look after yourself carefully whilst reading it. I cannot stress that enough. Before you start, identify at least two people you could have a chat with, on the phone or over a cup of tea, if you start to feel distressed while reading. If you cannot identify anyone, because you are feeling totally isolated, then in the UK you could call the Samaritans on 08457 90 90 90 or you could email the Spiritual Crisis Network at info@SpiritualCrisisNetwork.org.uk. Elsewhere you can phone or email your country's Spiritual Emergence Network, if there is one (see Resources at the end of the book).

A good idea, if you start to feel upset while reading is to put the book down, for the time being, and come back to it later. Focus on what you need, on listening to your body, on trusting your instincts. Focus on taking care of yourself, nurturing and nourishing yourself and getting support.

Even if you are not going through spiritual emergency right now, you may still find parts of the book disturbing. Some of the personal stories are not easy to read. However long ago you went through your crisis, aspects of it may get activated by reading this material. This may be the case, for instance, if you were not able to fully integrate the experience at the time. Again, please take care, look after yourself and do not hesitate to reach out to friends or family for support.

Introduction

The flame of the candle flickers. Mirrored in the glass of the tea-light holder, it looks like two flames dancing a duet. But I know that to be an illusion; there is only one flame. The world we live in is like this. There is only one whole and yet we firmly cling to the illusion of duality, that everything is separate. This is one of the insights that came to me experientially when I was going through spiritual emergency in 2006.

Spiritual emergency often brings insight with it, usually at the experiential level, insight that we cannot put into words, insight into ineffable truths, a knowing that words can only point to. It gives us a glimpse, and sometimes more, of ultimate Realization. This glimpse points us in the right direction, so that we know what to look for on the spiritual path. How can we search for gold if we have never seen it and have no idea what it looks like? Spiritual emergency opens the door, maybe only just ajar, but enough for us to move towards fulfilling our full spiritual potential. This, for me, is the most valuable aspect of spiritual emergency. What in the world could be more precious?

These glimpses need nurturing and cherishing. They will lead us where we want to go. The challenge with spiritual emergency is that they can get totally lost amongst the traumas of the crisis. For traumatic it undoubtedly is. There is no escaping the horrors of spiritual emergency when it grips us. It deals in extremes; our greatest blessing and our worst nightmare, all neatly packaged up into one single bundle; heaven and hell co-habiting.

Spiritual emergency is a crisis of spiritual awakening. It can happen to you if you are actively engaged on a spiritual path, but it can just as likely come seemingly out of the blue, even if you do not think of yourself as particularly 'spiritual'. It is a spiritual awakening that has speeded up into an unimaginably intense state that is difficult to manage. The 'dark night of the soul' and what is sometimes called 'mystical psychosis' are both forms that spiritual emergency can take.

My sincere hope in writing this book is that you will come to appreciate how vitally important spiritual emergency is; how much it has to offer us collectively, if properly understood and supported. It is largely unknown, one of those well-kept secrets. Now, as we move towards 2012 and beyond, is the time for it to become more universally recognized.

Why? Because we are experiencing a considerable acceleration of the spiritual evolution of mankind. When the process of spiritual awakening starts to speed up, it often tips over into spiritual emergency. It becomes too much for us to cope with and presents us with enormous challenges and very real dangers. Not only is the actual crisis itself extraordinarily difficult to cope with, but once we have got through that we have to renegotiate our place in the world. If we are aware of spiritual emergency, either personally or professionally, we stand a much better chance of turning the very real dangers into an opportunity for healing, growth and spiritual fulfillment, for ourselves, for our clients and patients.

In Case of Spiritual Emergency will provide you with a map to find your way around these psychospiritual crises. In Part I, you will find out what it can be like when the process of spiritual awakening gets out of hand, when a person can no longer function at an everyday level. You can read about the intense physical, emotional and energetic features of spiritual emergency. I draw on the personal experience of those whose spiritual unfolding led to crisis. These are ordinary people, like you and I, who have navigated their way through the terror and bliss that can accompany spiritual awakening. Their stories show how they coped and what they learnt, as well as what can trigger such crises. You will also find chapters which show spiritual emergency's place in Western psychology and which survey spiritual crisis related research. These help show the scientific and medical validity of the phenomenon.

In Part II, you will read about individuals, down through the ages, whose spiritual awakening has been accompanied by spiritual emergency. These are the mystics, the creatives, the present day figures. I want to give you a sense of how spiritual emergency cuts across time, cultures and spiritual traditions. It is not a recent occurrence by any means. Ever since mankind has been awakening spiritually we have been experiencing spiritual emergency.

In Part III, you will find the Three Key Phases of Moving Successfully Through Spiritual Emergency. Phase 1 involves *Coping with the Crisis*. A central tool here is Mindfulness. I believe it to be the linchpin in coping with spiritual emergency, especially for becoming grounded and working with the fear. There is also a very practical chapter covering issues such as getting support and looking after our physical needs.

Phase 2 will guide you in *Making Sense of It All: Using the Hero's Journey to integrate the experience*. The Hero's Journey is a powerful model that will help you understand what you have experienced. It stems from the work of the world-renowned mythologist Joseph Campbell. Through the medium of age-old stories and legends,

The Hero's Journey draws on archetypal symbolism and imagery to chart the struggles and victories of what it means to be human. It is an enormously versatile model, so much so that it will be relevant for you whatever you have experienced.

Phase 3 of Moving Successfully Through Spiritual Emergency looks at how to cope with the challenge of *Going Back Out into the World*. We will not be able to carry on living our lives as before, as if nothing has happened. What changes do we need to make? How can we bring back the hard-won Elixir, the reward of the Hero's Journey?

We can do this once we have dealt with the worst of the crisis and have started to make sense of it. In reality the process is not quite as linear, not quite as clear-cut as moving from Phase 1 to Phase 2 and on to 3. The process is more organic and we may move in and out of the different Phases or they may overlap.

In this last part of the book, Moving Successfully Through Spiritual Emergency, there are questions for reflection and points of practical guidance, in easy-to-use boxes. If you are reading this book because you personally are going through or recovering from spiritual emergency, it is a good idea to use a notebook or journal and work through these suggestions. Then *In Case of Spiritual Emergency* will not be just another book to read, but a potent healing process to actively engage with. If you are reading this book as a supporting friend, carer or professional you will learn plenty to guide you, to help you be the best support you can.

After the End Word, which looks at the role crisis plays in global awakening, there is a section on Resources. This lists many different websites, DVDs and places to stay when recovering from crisis. Not only will you have gleaned much practical guidance from the Three Key Phases of Moving Successfully Through Spiritual Emergency, you will also have an invaluable reference tool to take off the shelf whenever you need it.

PART I

What is
Spiritual Emergency?

The Unbroken

There is a brokenness
out of which comes the unbroken,
a shatteredness
out of which blooms the unshatterable.

There is a sorrow beyond all grief
which leads to joy
and a fragility
out of whose depths emerges strength.

There is a hollow space
too vast for words
through which we pass with each loss,
out of whose darkness we are sanctioned into being.

There is a cry
deeper than all sound
whose serrated edges cut the heart
as we break open
to the place inside,
which is unbreakable and whole,
while learning to sing.

RASHANI RÉA

CHAPTER 1

A Crisis of Spiritual Awakening

THE AWAKENING OF GLOBAL CONSCIOUSNESS

On a clear April day, I stood on top of Magdalene Hill, high up, overlooking Turin, with the snow-capped Alps in the background. Take a moment to picture the scene; the warmth of the sun, a slight breeze, the majesty of the mountains. I was at the tail end of a terrifying and powerful period in my life. Any memories of how desperate I had been at the height of my crisis, on the verge of killing myself, were becoming just that, memories.

It was no accident that my crisis had taken me back to Italy. I had spent part of my childhood growing up in Tuscany, climbing olive trees and picking up Italian at the local junior school. During that intense month-long spiritual emergency in 2006, I felt the invaluable support of an Italian spiritual teacher. I had never met him, but a close friend had given me a photo of this man. Now I was very grateful. I felt drawn to Italy as a place where I could feel safe.

As I stood on top of Magdalene Hill on that warm spring day, the Italian teacher came through to me. Psychically, I was still in a very open state. I felt his sadness as he said, in Italian, 'Look how beautiful she is', meaning the earth. 'And we have got to leave her'. Did he mean when we die? Or did he mean the human race as a whole will have to leave this stunning planet? I do not really know, but when I then read Eckhart Tolle's *A New Earth*, his words struck a chord. Our choice as a species is simple, he tells us: evolve or die. He puts it urgently and starkly. He is talking about evolving spiritually.

If we are going to 'wake up', individually and collectively, in sufficient numbers, we need to be familiar with the process of spiritual awakening. We need to be aware of the potential dangers, so that if our journey becomes challenging, we can deal with it successfully. Given the unprecedented growth in the level of consciousness that is taking place, knowledge about spiritual emergency is all the more important. There is no doubt that being familiar with the challenges that can arise during spiritual transformation helped me to cope.

As the awakening of human consciousness gathers pace, more people are having to navigate the treacherous waters of spiritual emergency. As more people succeed in finding their way through such experiences, their awakened consciousness, in turn, helps the planetary shift to gather further momentum.

As we explore and become familiar with the territory of spiritual awakening and spiritual emergency, we realize that there are very real parallels between what happens at the individual level and what is happening at the global level. We are beginning to experience global crisis; the awakening of global consciousness is starting to feel very painful and challenging. This is no different from what individuals experience when going through the birth pangs of spiritual crisis. The lessons we learn at the personal level can be translated to the global.

Humanity is embarked on its own version of the Hero's Journey. If we understand the process at the individual level, we can know that the seeming breakdown of what I call the 'dark night of the globe', is the precursor to breaking through to a whole new level of awakened consciousness. We will see just how intense and challenging a crisis of spiritual awakening can be personally. It is this level of intensity and challenge we can expect to see at the world level during the coming years of transition. When we read the stories of those whose spiritual unfolding led to crisis and see how they navigated their way successfully through it, we can hopefully draw much inspiration and encouragement from that in responding to the global situation.

Spiritual awakening

So what is spiritual awakening? It is above all a process; a process of exploration and unfolding; a process of learning and growth, of healing and purification. It involves the whole of our beings and works on all levels, physical, emotional and psychological, as well as spiritual.

For me, as for many, it has been a journey of self-discovery, of learning to love myself. In our 21st Century mind, body and spirit culture that sounds like such a cliché; but I had been brought up with years and years of violent verbal abuse, criticism and bullying from an alcoholic father. Learning to love myself was far from a cliché. It was hard work. Yet the quality of our life and our relationships vitally depends on it if we are not going to carry on endlessly recreating the cycle of being abused.

So it is also a journey of healing past trauma and wounding. That in itself could be done in a very secular way, through counseling, psychotherapy and personal development. What is it that makes it a clearly spiritual path? Personally, choosing work that enables me to put my spiritual practice at the heart of my life has been key. I now teach meditation and Mindfulness, a practice which is with us throughout each day. I find going on retreat, especially silent ones, to be powerful and deeply nourishing. A quiet, simple lifestyle is essential for me too.

Along the way, the qualities that naturally develop are those of trust, faith, gratitude and devotion. We come to a place where we can stand fairly and squarely in our power, the strength of which is coupled with the softness and receptivity of surrender. And it is our trust in the Universe, Source or God that enables us to surrender. Above all, we learn to trust our direct experience, to nurture the glimpses of awakening that we have, whatever the circumstances in which they arise and whatever the verdict of mainstream psychiatry.

WHAT IS SPIRITUAL EMERGENCY?

The natural process of spiritual unfolding I have just described can be gentle, gradual, even graceful. I would like to be able to say that it is usually so, but the more I explore this field, the more people I talk with, the more convinced I become that very few escape some sort of crisis, critical choice or dramatic turning point. Whether that turns into spiritual emergency depends on its intensity. Spiritual emergency, a term first coined by Stanislav and Christina Grof, is essentially an intensifying of the process of spiritual awakening, a speeding up of the process that becomes unmanageable and often terrifying.

Whereas we might prefer our spiritual growth to be like a gentle paddle down the stream, spiritual emergency is more like the rough ride of a speed boat at full throttle. As it involves psychological transformation as well as spiritual, it is sometimes known as psychospiritual crisis. One way of seeing spiritual emergency, therefore, is as a complication of our natural development as spiritual beings.

Another way of approaching spiritual emergency is from the angle of mystical experiences. In a mystical experience you might feel waves of bliss and awe; you might lose your sense of an egoic self, experience the Divine and much more.[1] These are experiences beyond words, beyond the grasp of the intellectual mind. Spiritual emergency often includes many of these elements.

For many of us, because this spiritual transformation also involves psychological transformation, when we start to open up to the transcendental, any unresolved aspects of our personality can come to the surface. All the wounding we carry, any trauma we have ever been through, any parts of ourselves we have repressed, known as the 'shadow'; all these and more can come up, demanding attention, asking to be healed, resolved. Then what we experience is more likely to look like the spiritual emergency the Grofs identified. We may find it difficult, if not impossible, to cope with everyday life. Basics tasks of looking after ourselves, like cooking or even washing, may feel too much. Our inner world may take over, merging confusingly with the outer world.

There is a third approach to understanding spiritual emergency. In recent decades, efforts have focused on 'psychosis' and mental health issues, to try to distinguish spiritual emergency from these. This is a complex issue, which we will look at in more detail in Chapters 3 and 4. My personal take on it, and this is also the stance of others in the Spiritual Crisis Network, is that I am not interested in trying to distinguish between so-called psychosis and spiritual emergency. I take the view that it is all the psyche's attempt to heal and move towards wholeness, that each experience is potentially spiritually transformative.

This is informed by my own experience, which is common to many others. In 1996, I was overwhelmed by stress, coping with the aftermath of separating from my husband and dealing with a very difficult working relationship with my boss. All I wanted was some help in coping with my stress levels that were rapidly getting out of hand. I joined a meditation class and bought a book, an introduction to meditation. The final chapter described some of the states and feelings meditators aspire to after years of practice, feelings of oneness and interconnectedness with all things, of unbounded love, of blissful inner peace.

I instantly recognized such states. 'I've experienced that,' I thought. 'I know what that feels like.' But there was no mention in the book that such states could be followed by apparent mental breakdown, a month in a psychiatric hospital and losing a year of one's life, as I had experienced at the age of 20. Suddenly I knew I was on to something. This was confirmation of what I had known all along, that there was far more to my so-called 'breakdown' than met the eye.

So it is not that simple, trying to say 'this person is psychotic, whereas this person is going through spiritual emergency' as the two so often go hand in hand and cannot easily be separated out. It is more a question of both being present, rather than either/or.

All of these ways of seeing spiritual crisis bear some truth. Spiritual emergency can certainly be experienced as a naturally healing process that takes us towards greater wholeness; an evolution towards fulfilling our true potential as spiritual beings. It also contains elements of mystical experiences, such as feelings of oneness with the Universe, *and* it often contains psychotic-type elements. This combination of the mystical and the psychotic-looking means that some cases of spiritual emergency take the form of what we might call 'mystical psychosis'. Alternatively, it can take more the guise of depression, of a 'dark night of the soul'. It is as if the powerful energies that we have to contend with can either take us up into the heights or down to the depths.

DANGER OR OPPORTUNITY?

The Chinese symbol for 'crisis' consists of two characters, one that denotes 'danger' and the other 'opportunity'. A crisis of spiritual awakening holds both of these. There are, without doubt, some very serious dangers. There is, however, nothing 'wrong' with spiritual crisis in itself. It offers phenomenal potential for spiritual growth and healing. In order to fulfill the opportunities, in order to experience such a crisis as the wonderful gift and blessing that it can be, we need to be fully aware of the dangers. The more conversant we are with these the more likely we are to be able to safeguard against them.

DANGERS

Tragedy

The dangers associated with spiritual emergency are not to be underestimated. In rare and extreme cases it can end in devastating tragedy. People can die, either because of accidents or because, in their despair, they take their own lives. For Marie Moore, whose story you will read later, the torment she was going through became too much to bear and she shot, first her son, and then herself.

One specific danger here is that it is possible to become so out of touch with one's physical body and this material realm that someone thinks, for example, that they can fly or breathe under water. It is easy to see how someone could harm, or even kill themselves by accident, in such circumstances.

> AARON was an exceptionally warm-hearted and gifted 17-year old and, by all accounts, a very spiritually mature soul. After a series of traumatic events he went through spiritual emergency and a tragic set of circumstances unfolded. At a crisis point, whilst with his mother, he phoned the police for help. When the police arrived they sectioned (committed) Aaron under the Mental Health Act. He was kept in a cell overnight, pending psychiatric assessment, with no contact whatsoever allowed with his family, despite how young he was. He was then transported alone and handcuffed in the back of a police van to a psychiatric hospital. A journey that should have taken one and a half hours took three, because the driver got lost. Later, at the hospital, when Aaron sat down to chant in the way he had grown up with, he was told to go to his room. He pleaded with staff, 'You must respect my faith'.
>
> The next day Aaron's rights under the Mental Health Act were read to him, explaining that being sectioned meant he did not have the right

to leave the ward. No member of his family was present while the document was read to him. With typically strong determination, he ran out into the garden and easily scaled a high fence, running as fast as he could towards home. He was found on some wasteland having already sustained fatal head wounds by bashing his head against a discarded gas canister.

Aaron's story highlights how very important it is for someone experiencing such intense times to be kept safe. He had told staff that he needed to express the intensity of the emotion he was feeling in his body and was concerned that he might unintentionally hurt himself, that he only needed to be kept safe. He had told his mother 'all I need is a padded cell'.

Despite the depth of her grief, Aaron's mother tells beautiful stories of the profound spiritual experiences she always had of her son and the especially numerous mystical events surrounding his death. She feels very strongly that this was his path.

Hospital

It is common, unfortunately, for people going through spiritual crisis to end up on psychiatric wards, not because that is the best place for them in their sensitive state, far from it, but because there are so few places that can offer the level of holding they need. Such a person often needs intensive 24-hour care.

There are several dangers associated with ending up in hospital for someone going through spiritual crisis. One woman, sensing these, made a run for it, knocked on somebody's front door, saying that her car had broken down, not herself, and could she please call for a taxi. She made her escape. Her fate was very different from Aaron's.

Crisis pathologized as illness

The whole relationship between the symptoms of mental distress and those of spiritual crisis is phenomenally complex. We need here to be aware of the unhelpful tendency to think in either/or terms, that a person is either psychotic or going through spiritual crisis, as I mentioned earlier. The reality is that very often psychotic-type elements go hand in hand with the spiritual emergency and the experience is far more likely to be a question of both/and. The danger is, however, that mental health professionals will not understand this and, because of the psychotic-type elements, will tend to pathologize the entire process as illness.

I cannot stress too strongly how damaging this is. There are literally thousands of people who have been through the mental health system who have not had the spiritual aspect of their experience honored. The spiritual dimension has been completely overshadowed by the interpretation given to their experience by the medical model. I know this first hand. When I became concerned about my mental health, after my husband and I split up, I asked my doctor to refer me to a psychiatrist. As we discussed possible preventative measures, I remember trying to explain to her about unitive or mystical states of consciousness. Her take on it was that this was part of the early stages of 'illness'. She was totally dismissive of the idea that what I had previously been through might in any way have been a spiritual experience, had a spiritual dimension to it or be of any value.

Part of why this is so damaging is that the spiritual aspect of our experience is what can nourish and support us after we have been through such a cruelly bruising time. Not to mention the lost opportunity for healing and moving towards integration and wholeness.

Unfortunately, what tends to happen is that many of us buy into the pathologizing perspective of the medical model ourselves, if we do not have an alternative framework with which to understand what is happening to us. If those around us repeatedly tell us we are ill, in our vulnerable, impressionable state, we end up believing them. So the dangers of lack of understanding can apply equally to us as individuals as to health professionals.

> When Aaron arrived in the police van his mother was waiting for him at the hospital entrance. As Aaron put his handcuffed hands over his mother's head to hug her he said, 'I didn't realize I was this mad'.

Over-use of medication

The second very real danger of finding oneself in hospital is that the process, which is essentially one of healing, can be completely interrupted by the trauma of hospitalization. Part of the problem is that the level of medication used for someone going through spiritual crisis tends to be far too high. Prof David Lukoff, an authority in this field, stresses the need to evaluate for medication, with a view to keeping it to a minimum. A doctor interviewed in Kaia Nightingale's documentary film *Spiritual Emergency* explains that the doses needed for such people are almost homeopathic they are so small.

This is not to say that medication does not have its place. It can be helpful at times. A related issue is the amount of choice that a person has over whether to take it or not. Annabel's experience of being forcibly medicated is, unfortunately, not uncommon.

> "I was committed to hospital (sectioned), and injected several times with high dose anti-psychotic medication against my will, which was

deeply traumatic... Over the course of the next four years I was hospitalized and sectioned two more times. During these years I received valuable support. One psychiatrist stands out particularly because she honored the spiritual dimension, whilst also realizing the benefits of medication when used sparingly."

ANNABEL

This needs to be balanced with the fact that some people have a positive experience of being in hospital. A woman who is a natural health practitioner was fortunate that her mother managed to persuade staff that to medicate her daughter would go totally against all that her daughter believed in and worked for professionally. Consequently this woman had a relatively good experience of being in hospital.

Somebody else who was feeling deeply suicidal and did not feel safe at home alone, having already made several suicide attempts, found hospital to be a reassuring and relatively safe place, something of a safe haven.

Overall, perhaps the greatest danger of ending up in hospital, and certainly the saddest aspect, is that the opportunity for healing and growth, for living a fuller, richer, more awakened life, can be irretrievably lost. The natural process of renewal, as the psychiatrist John Weir Perry called it, can be totally thwarted. Both the trauma of hospitalization and the over-use of medication can have this effect. And once the process has been stopped in its tracks it can be difficult, if not impossible, to retrieve.

Unable to work

Not being able to function at an everyday level when going through spiritual emergency has some very tangible consequences. Depending how long the crisis lasts we may lose our jobs or have to give them up. We may also lose our homes.

"As the 'other worlds' experiences slowed down, so I began to feel incredibly unwell physically. This was the next phase. During this period my partner and I lost our home due to my not working and his redundancy. We moved in with family. I was feeling increasing pressure to get myself together and get a job again. Eventually we relocated to a city hundreds of miles away and I began the task of knocking on doors and applying for jobs. All the while I was screaming inside and exhausted. I was drinking more coffee and smoking more cigarettes in an attempt to keep going. Finally when I did get a job I collapsed in the ladies bathroom after a few weeks."

KIMBERLEY

In the longer term, many find it difficult to meet their material needs, to earn a living, either because the crisis itself drags on over years or because they struggle to integrate back into everyday reality.

Isolation

We may become estranged from family and friends, who cannot understand what we are going through. All this, along with probably having to stop working, can result in our feeling extremely isolated. It can be all too easy, therefore, when we are going through spiritual crisis, to feel very alone with the experience, especially if we do not realize it is a recognized phenomenon. This sense of isolation is something that many who have been through such experiences speak of. Often it is only afterwards that people discover others have been through it too.

We have seen that there are many very real dangers associated with spiritual emergency. It needs handling extremely carefully and sensitively because of these. It is worth stressing again, however, that there is nothing wrong with spiritual crisis in itself. I have yet to come across a single person who regrets what they have been through. Even Aaron's mother has an awareness of the bigger picture, of it being his time to leave and of his blissful existence in another realm, which continues to be a deep contribution to her own spiritual journey.

OPPORTUNITY

The transformation that comes with spiritual emergence and emergency is potentially radical. Whatever the degree of awakening we have experienced there will be the opportunity to lead our life on a completely different level. We are likely to experience greater quality of life, living deeper, richer lives, more full of meaning and purpose than before. Psychologically and emotionally we are likely to be more fully integrated, to be more resilient to the ups and downs of life and to meet challenges with greater composure. There may be a new-found sense of wholeness.

> "My experience was a wake-up call to my own deep imbalance and need for healing. I know I've arrived much more fully into my body, so that life can now unfold in a much more fruitful way than before."
>
> *ANNABEL*

A crisis of spiritual awakening offers far more than simply the opportunity for psychological healing and growth, which is not insignificant in itself. If the crisis came out of the blue and we were not originally engaged in any spiritual exploration, then this experience will set us firmly on a spiritual path. If we were already actively pursuing our spiritual development, then the learning that comes with such an opening is likely to be truly profound. It is an experiential learning that no amount of reading books can replace. Much of the spiritual wisdom we will have gained cannot, in fact, be expressed in language. With the awakening there comes a certain knowing, a confidence and a trusting. As I wrote in the UK publication *Caduceus,* the experience continues to nurture and nourish me in profound ways every day.

All of this can lead to a deep sense of gratitude, of finding joy in the simplest of things. So often, when a person has been through a crisis of spiritual awakening, afterwards they have a very strong urge to be of service, to give something back to the world.

> "As my life now moves forward from my 40th year my desire is to be of service to life and to uncover and to utilize my talents and gifts to benefit the world. This like all worthwhile processes will take time and be a step-by-step unfolding. I hope now to be in a position to be a support to others as they travel over disturbing, frightening territory on their way back to wholeness. Offering talks about my healing journey has given me an opportunity to share the fruits of my experience with others."
>
> *ANNABEL*

CHAPTER 2

What Does Spiritual Emergency Look Like?

Spiritual emergency is not a new phenomenon. The process of spiritual development and awakening has never been easy. Both the Buddha, meditating under the Bodhi tree, and Jesus, in the desert, faced their own crisis points or 'dark nights of the soul'. So did many of the Christian mystics, such as St. Teresa of Ávila and St. John of the Cross, who gave us the term 'dark night of the soul'.

Spiritual emergency, or crisis, is coming to more people's attention now because we are living in unprecedented times. The stories you will find in these pages represent the modern day equivalent, as more people are experiencing various degrees of awakening. It is difficult to describe how intense spiritual emergencies can be. We can feel utterly helpless, overwhelmed by its suddenness or its intensity, by the inrush and upsurge of spiritual energies.

In this chapter we will cover some of the key features, with descriptions from Kimberley, Jennifer, Kate, Emma, Annabel and myself of what it was like for us.

KEY FEATURES OF SPIRITUAL EMERGENCY

- The intensity of the experience can consume our whole being
- We can find it impossible to cope at an everyday level
- Our inner world can take over and blur confusingly with the outer world
- We can have unusual physical pains and sensations and find it impossible to sleep
- We can experience a rollercoaster of powerful emotions
- There can be a sense of everything falling away, including our sense of self in an ego-death, albeit usually temporarily
- There may be ego-inflation; for example, we may believe that we are the reincarnation of Jesus or Mary

- Thinking can become confused as the rational mind desperately tries to make sense of what is going on, resulting in psychotic-type elements
- Symbolism and mythological themes become very meaningful for us, for example, sacred union or marriage, the battle of good versus evil, etc.
- Meaningful coincidences, known as synchronicity, often become more frequent
- We might see unusual things, such as past-life flashbacks in our mind's eye or spirits
- We can experience sudden and strong energies, either as our own life force energy is spontaneously released or with the powerful inrush of spiritual energies

"The most extreme phase of my experiences lasted about five months. Every night I had 'seizures' with heat, convulsions and cramps moving through my body. Each time accompanied by voices, images flashing at high speed in front of my eyes whether open or closed, swirls of colour swimming around the room, nausea, body pains and exhaustion."

KIMBERLEY

THE BODY

The process of spiritual unfolding is essentially one of healing, growth and purification. This takes place on all levels, physical, emotional, psychological and spiritual. At the physical level, the body holds a great deal. Any shock, trauma, emotional or psychological wounding is held at the cellular level. It is as if the body has a memory and it is stored in our tissues, muscles and organs.

If our spiritual journey progresses smoothly and gradually, there may well be moments of healing which include the release of old trauma. A massage or psychotherapy session, for instance, might release something. As the body releases we may experience temporary shaking, as a gazelle would in the wild after a close shave with a lion.

Afterwards our body may feel positively different in some way. We may feel a new openness across the chest, where before it felt tight. A physical complaint may clear up. The possibilities are endless. Usually, however, this process of healing and spiritual growth is slow enough and gentle enough for us to easily cope with any physical aspects. We may get temporarily thrown off balance, we may even need to take a short period of time off work, but overall the process is manageable.

Sometimes though, for a variety of different reasons, the process speeds up. We may suddenly touch on particularly deep childhood trauma or we may have made a new, stronger commitment to our spiritual path. We might go on a week-long yoga or Tai Chi course, which intensifies our practice. The potential triggers are many and we will look at them in detail at the end of this chapter; and sometimes it is simply not clear why the process of spiritual development suddenly speeds up. What becomes clear is that it is too much for us to cope with. Things start to get out of hand; the experience consumes our whole being.

Physically, when we tip over into spiritual emergency, we can feel a whole range of unusual pains and sensations. Some get unexplained spasms and jerking. Very commonly people have feelings of vibration in the body. Frequently it is difficult or impossible to sleep, sometimes for weeks on end. You might have digestive problems too. Many go to the doctor, who invariably cannot find any obvious physical cause. Jennifer experienced particularly extreme physical effects when going through spiritual emergency.

"Physically, I experienced pain so intense that there is no description for it. I became sensitive to everything. My immune system got overworked and confused. I became allergic to many things."

JENNIFER

Some people get burning sensations of searing heat in part of the body, or all over. Or you might get sensations of something crawling over your skin, or excessive wind, sometimes called 'holy belching'. Some hear a rushing or roaring sound in their head. When I went through spiritual emergency in 2003, I found myself travelling back to the UK from Egypt in a wheelchair. What I had been through was so extreme, so overwhelming emotionally, psychologically and spiritually, that my body had gone into shock. My legs had given way and I had not even been able to get from my hotel bed to the toilet. Here is just one aspect of Kate's experience:

"In a development that was to last the next ten years, my skin and nerves began manifesting the energy through prickling and tickling sensations in my hands, feet, arms, legs and 'electric ears'. The phenomena was so alarming at first that I was referred to a specialist to rule out anything serious."

These are just a few of the vast array of possible physical traits. They can range from unpleasant, painful or exhausting to completely debilitating. As the examples show, the physical effects of spiritual emergency alone can be extreme and very frightening. Many going through such experiences will, however, meet with much less severe physical effects, which can nevertheless be upsetting and challenging.

POWERFUL EMOTIONS

Any process of gradual spiritual awakening is likely to involve powerful emotions at times. Under normal circumstances we can contain these and although we might need emotional support from family, friends or a therapist, this is likely to be manageable for us.

In spiritual emergency, as things start to get out of hand, so the emotions can be completely overwhelming. We are likely to experience a whole range of powerful emotions.

> "I was opened up to the beauty in everything and it made me cry with joy."
>
> *EMMA*

Feelings of unbounded love, of sensitivity to the suffering of the world and of compassion are all common. Such strong feelings of love often go hand in hand with feelings of total calm and peace.

Whilst these are obviously positive, what can be difficult to cope with is flipping from such states into, for example, strong feelings of fear. It is easy to see how these powerful extremes of emotions, from peace and love to terror, can feel like an emotional rollercoaster. The speed with which we can flip from one state into another only adds to the sense of being out of control. During my brief stay in Egypt, at one point I experienced what felt like the entire suffering of humanity going back over thousands of years. Within just a few hours, I also experienced our natural state of inner stillness, of a spaciousness that has no physical boundaries.

Fear is by far the most common emotion felt during spiritual emergency. It can range from anxiety, dread or panic to abject terror. These feelings can be caused by the disturbing physical sensations already mentioned, by fragments of past lives coming up spontaneously, or by any number of the features of spiritual emergency.

> "I was terrified to go to sleep because this was when the seizures would happen. I was terrified to be awake because I could see and hear spirits walking around my home. I could feel no emotion but fear. This wasn't any fear I had felt before or have felt since. It was a realization that I was no longer safe in my own body or mind."
>
> *KIMBERLEY*

No longer feeling safe in one's own body or mind is something that many who go through spiritual emergency will identify with. It is far more terrifying than feeling threatened by someone or something external.

It is not surprising that when going through such powerful spiritual and psychological transformation people often fear they are going mad. The relationship between mysticism and madness is a complex one, summed up by the Indian term 'holy madness'. Jennifer, the psychologist, became fearful of going mad.

> "At one point in 1993, I became fearful to my core about the possibilities of evil or psychosis; and, for about three weeks, I could not feel the presence of God. The best descriptions of what that felt like are in the Biblical descriptions of hell. I was lost and bereft with no anchors. Fear took over."
>
> *JENNIFER*

Another common fear is of dying. On more than one occasion, Emma felt scared that she was going to die and Jennifer felt her life threatened by the extreme physical effects of her crisis.

> "I remember thinking I was dying. All that kept going around in my head was the thought that this was death. I was experiencing death, hearing it, feeling it, smelling it and I was fighting for my life."
>
> *KIMBERLEY*

EGO DEATH

Ultimately, through the process of spiritual development, we are seeking to transcend the ego, the limited sense of self. This can, however, be a terrifying process for the ego, and for us, because we are so identified with the ego. As we start to move into transpersonal levels of consciousness, the ego senses there is no place for it. The fear of dying that many experience when going through spiritual emergency is largely the ego's fear of being extinguished.

It is very easy, in such extreme states, to confuse psychological or ego death with physical death. Hence the danger of physically dying can feel all too real and we can become preoccupied with our own death.

> "I asked my Dad how he knew that I wasn't dead. He said that was easy. He knew that I wasn't dead because dead people didn't need to eat.

Simple, logical answers like this were very reassuring."

EMMA

As Stanislav Grof says, "During the ego death…everything that one is or was - all relationships and reference points… collapse, and the person is left naked, with nothing but the core of his or her being".

As we have seen, in very rare cases, a person moves so far beyond ego and the physical body and loses touch with this everyday level of reality to such an extent, that they take their own lives by mistake.

EGO INFLATION

Many spiritual seekers will at some point experience a degree of pride. Through their spiritual practice and with self-awareness, they may see this for what it is and consciously work with it. The inflation that can happen during spiritual emergency is somewhat different. Because the whole process is speeded up, there may be relatively little self-awareness, with little space around what is happening.

In ego inflation, the ego appropriates to itself the powerful spiritual energy that is coming through. Roberto Assagioli, the founder of Psychosynthesis, a spiritual approach to psychotherapy, explains this confusion of the higher, spiritual Self and the egoic, personal self.

As the personal self experiences the spiritual Self, the individual has a sense of greatness and expansion, and feels as if they are part of divine nature. It is not that we *are* Jesus, for example, but that we are experiencing the energy of the Christ Consciousness. The ego and the mind are not used to such extraordinary events and get confused. So, whilst …

> "…the inflowing spiritual energies may have the unfortunate effect of feeding and inflating the personal ego…instances of such confusion… are not uncommon among people dazzled by contact with truths which are too powerful for their mental capacities to grasp and assimilate."
>
> *ASSAGIOLI* [2]

Understanding this means that we can be compassionate towards ourselves and others at such times. After I had experienced the energy or consciousness of the Virgin Mary in this way in Egypt, my therapist wisely encouraged me to explore that contact and what it meant to me, rather than merely dismissing it.

PSYCHOTIC-TYPE ELEMENTS

Ego inflation is known within psychiatry as 'grandiosity' or 'delusions of grandeur' and is often considered to be a classic symptom of psychosis. This is a great shame, given that, as I have just described, something very significant is likely to be happening for the individual. In the same way 'visions' are likely to be seen as 'hallucinations' by mainstream mental health professionals. We ourselves can become confused and not trust our experience.

> "I had a Shiatsu treatment that opened me up spiritually. When I got home, I put on a piece of music and the lyrics took me deeper. I began to dance whilst dusting to the music and then a pure, liquid like, golden ribbon of light came from my navel area. It was attached to the duster handle and the ribbon fell into beautiful folds. Then I had a thought that I might be hallucinating and immediately the ribbon disappeared."
>
> *EMMA*

Also, when going through spiritual emergency, our thinking can become confused, as the rational mind desperately tries to make sense of what on earth is going on. So the difficulty with some of the features of spiritual emergency is that they can easily look like psychosis. The important thing is not to throw the baby out with the bath water.

> "I don't want to get into the debate about whether something is a spiritual emergency or a psychotic episode, though the former is more palatable and dignified. What matters to me is that my experiences have opened up my mind to the possibility of there being more than our human senses can detect."
>
> *EMMA*

Spiritual emergency can also look like depression or 'bipolar disorders', which used to be called 'manic-depression'. Kate puts it very well:

> "I've had many years to reflect on my process. In doing so, I've become aware that my experiences mirror Bipolar Disorder II in certain ways and that certain professionals would diagnose me as such. Along with many others, I disagree with the imposition of biomedical illness models upon such profound and life-changing psycho-spiritual experiences and believe that extreme mood shifts are an intrinsic part of the cleansing process."

MYTHOLOGICAL THEMES

The process of spiritual awakening takes us into transcendental territory, into the realm of archetypes and symbolism. Jung considered that to think literally is to inhabit the material world; to think symbolically is to inhabit the spiritual.

When we find ourselves catapulted into spiritual emergency it is as if we have no choice. As we move into our inner worlds and temporarily cease to be able to function in the outer world, archetypal energies take center stage almost against our will. Mythological themes take on great significance for us, whether of sacred union or marriage, the uniting of the masculine and feminine principles or the battle of good versus evil. The archetypal image that came to me was that of Avalokitesvara, the Buddhist Bodhisattva figure who wept for the sorrow of mankind and who has a thousand arms to help relieve all the suffering. For Kate it was Persephone:

> "Unaware of the Greek myth of Persephone, I developed a mysterious obsession with pomegranates from the corner shop next to my lodgings and was eating one a day. As the urge to consume the blood-red seeds Persephone ate before her abduction by Hades faded, the narrow doorway to the Underworld shut behind me."

We have seen that the process of dissolving the ego, however temporarily, can bring up fear of death. And perhaps this is the most compelling of all themes in spiritual emergency, the universal theme of death and rebirth. The process of psychological death and rebirth can be mirrored in fears about the end of the world. A person might have terrifying visions of Armageddon.

Equally, the inner processes of the psyche might reflect outer symbols of rebirth. Someone might find a snake's skin that has been shed on a moorland walk, or they might come across eggs, anywhere and everywhere, including the most unexpected of places.

SYNCHRONICITY

The word synchronicity refers to meaningful coincidences, coincidences that can jolt us awake because they seem so unlikely or bizarre. When they happen, we get a sense of the bigger picture, of something greater than ourselves at work. Lindsay Clarke, the British author, once remarked that he gauges his psychic health by the flow of synchronicity in his life. When such coincidences happen there can be a sense of being in tune with the transcendent forces of the Universe.

In spiritual emergency, when we are opening up to this transcendental realm, we often experience a lot more synchronicity than we might usually. It is as if we are more fully in tune with the other dimension of reality.

> "The final showdown was accompanied by incredibly intense synchronicity — a magnetic principle of mutual unconscious attraction first defined by Carl Jung that manifests through meaningful coincidences. Over the years, I've experienced its subtle (and not so subtle) workings in a myriad of ways, from the simple to the profound, the gentle to the life-shaking. The most powerful clusters have occurred around key times of breakdown and breakthrough.
>
> Many spiritual self-help books take the line of least resistance and claim following coincidences will lead you to your dreams. I say it's a bit more complex — they may drag you through the hell first for a confrontation with your deepest, darkest self and when you crawl out the other side, those dreams may have changed beyond recognition. My therapist gave me the best advice about synchronicity — pay attention."
>
> *KATE*

Denise Linn, author of *Signposts: How to interpret the coincidences and symbols in your life*, explains that at any moment we are walking in a veritable 'forest of symbols'. Under usual circumstances we filter most of these out and are not aware of them or their meaning. During a crisis of spiritual awakening, however, we become much more in tune with the symbolic, transcendental dimension. Whilst this can offer valuable guidance, the sheer volume can, in turn, add to the feeling of being overwhelmed that goes hand in hand with spiritual emergency.

VISIONS

Visionary experiences vary enormously and it is difficult to know how many ordinary folk have at some time or another experienced some sort of a vision. People tend to keep such things to themselves for fear of being ridiculed or labeled 'mad'. Kimberley saw spirits walking about her house.

> "I chose not to tell anyone the full extent of my experiences because I couldn't imagine why anyone would believe such things."
>
> *KIMBERLEY*

Marie Moore, whose tragic story you will read later, was tormented by visions of being buried alive and burnt at the stake. These kinds of images and flashbacks that start to appear in our mind's eye may be past-life related. They can be particularly frightening if we have not been brought up with the idea of reincarnation as part of our cultural belief system, unlike millions of others around the world for whom it is a given. Such images can seem like scenes from a film and they are often violent and disturbing. They may or may not relate to past-lives that we ourselves have had. When going through spiritual emergency we tend to be so open psychically that we can easily pick up on 'material' that is not our own, from what Jung called the 'collective unconscious'. Jennifer had many visions of a more gentle nature:

> "For years, I frequently had a vision of a giant image of Jesus in a very large wicker rocking chair, rocking on a beach by the ocean. He was there comforting me and holding me. A few years later, I had an image in my mind's eye of my trying to get into his lap to be comforted and he told me, I had grown too big to be in his lap anymore. I was growing spiritually. In the next vision, we sat next to each other on a bench by the ocean and talked."
>
> *JENNIFER*

Another type of vision might relate more to being shown our purpose in this lifetime, what we have incarnated for. All these different types of vision are important and need honoring and, in the case of disturbing past-life visions, healing.

UNUSUAL EXPERIENCES

As experiences go, visions are unusual. Many who go through spiritual emergency, however, report all manner of strange experiences that go far beyond visions.

> "Every time I walked past a mirror, I saw my face distort and change into the face of someone else, someone I didn't recognize but who somehow felt familiar to me... I could feel the consciousness or feelings of trees, flowers and animals... I would wake in the night, open my eyes to see my own reflection staring back at me in some convex mirror, only to realize that the mirror was the black, shiny pupil of a gigantic eye staring back at me."
>
> *KIMBERLEY*

Some of these experiences can be highly disturbing. Others, like Jennifer's, can be heart-warming and life-affirming.

> "Once I went to Florida for Thanksgiving week. One morning, just as the sun came up, I went for a walk on the beach. I was pondering a difficult relationship. As I walked, a heron flew over, landed, and walked by me, about six feet away, between me and the ocean. We walked together for about ten minutes. Then it flew away. As the heron went, I knew what to do about my problem."
>
> *JENNIFER*

STRONG ENERGIES

For those who are consciously engaged on a spiritual path there is usually a process of gradual opening to spiritual energies. On the whole these can be contained. The person's spiritual practice will deepen as they mature. What is alarming during spiritual emergency is the suddenness and intensity of the inrush of spiritual energies, which come unexpectedly, seemingly out of the blue.

Energy is central to our understanding of spiritual emergency, both in terms of these powerful spiritual energies and in terms of the energy meridians and the chakra system of our bodies. A chakra (the word comes from the Sanskrit, meaning 'wheel' or 'disk') is an energy center. According to Indian Vedic understanding, there are seven major chakras positioned through the body, from the base of the spine to the crown of the head. These make up the subtle energy system of our body and are responsible for our physical, emotional, psychological and spiritual well-being. The seven chakras are the root, sacral, solar plexus, heart, throat, third eye and crown.

Kundalini life force energy resides at the base of the spine and can be released spontaneously or through spiritual practices. As the powerful energy moves up through the chakras and the spine it can create the intense burning sensations mentioned earlier. This is because, as it does so, it comes across blocks in body tissue. This body armoring occurs where we are holding any emotional or psychological wounding and acts as resistance to the rising energy. This creates heat in the body, which can be very intense, as the Kundalini energy attempts to clear the blocks.[3]

> "It felt like the fire just burned through my centre. I ended up with some physical damage down the centre of my body. First, I had to have my gall bladder out. My stomach and lungs were weakened. I ended up with a bladder disease called Interstitial Cystitis, an inflammation of the

bladder wall. When I saw an x-ray of my bladder, it looked as if it was on fire."

JENNIFER

Taken for granted in Eastern healing methods such as acupuncture and Shiatsu, subtle energy has been virtually ignored in Western medicine. Yet our ability to ground the powerful energies coming through at times of spiritual emergency is essential to our wellbeing. As Jack Kornfield, the Buddhist teacher and writer, says, at times the energy "… can become very powerful, to the point where there is so much energy coursing through the body… The entire body will vibrate."[4]

The strength and health of the lower chakras in particular, the root, sacral and solar plexus chakras, are fundamental to our ability to be grounded, rooted in our bodies and on this earth. They are also essential to our ability to successfully assimilate the inflow of powerful spiritual energies during times of opening. We will look in detail at grounding in Part III, Moving Successfully Through Spiritual Emergency.

DURATION

How long an experience of spiritual emergency lasts can vary enormously. Through the Spiritual Crisis Network, we hear of both short-term and long-term spiritual emergencies. Some last days or weeks, others months or years, in some cases many years. Jennifer's story spanned over a decade.

Perhaps one of the longest spiritual emergencies on record is that of St. Teresa of Ávila. For twenty-five years she experienced strange, undiagnosed, painful seizures that often left her paralyzed. Eventually they gave way to ecstatic raptures and she went on to achieve remarkable things.

Kaia Nightingale, of the Canadian Spiritual Emergence Service, has done some valuable research in this area. She discovered that many people do not have just one experience of spiritual emergency, but several. This has been confirmed by our experience with the Spiritual Crisis Network. In other words, not only does the duration vary enormously, so too does the possible number of experiences. Emma's story is a good example of this, having experienced five different episodes. In fact, the American psychiatrist John Weir Perry suggests that where the process is interrupted by heavy use of psychiatric drugs or the trauma of hospitalization, and not allowed to take its natural healing course, then repeated episodes may well result.

"What I do feel sure about is that the experiences have a purpose and a meaning. If it is my soul calling me to a life that will serve me and the planet better, then I'm doing all I can to be part of the solution. If the process is thwarted because of the misguided help from the mental health services then I can look forward to as many more episodes as is required to do the finish properly."

EMMA

THE DARK NIGHT OF THE SOUL

Often associated with depression, the dark night of the soul is perhaps one of the most perilous forms spiritual emergency can take, because of the danger of suicide. Reaching out for support is imperative at such times. In the UK there is an excellent non-medical facility in London for anyone who is feeling suicidal, the Maytree Foundation. You can stay there for up to four nights, with someone available 24 hours, to help you through the worst time (see Resources). Alternatively, call the Samaritans on 08457 90 90 90 or, in America, the National Suicide Prevention Lifeline on 1-800-273-8255, both 24 hours a day. In Chapter 9, Coping with the Crisis, we will look in more detail at the vital question of getting support.

Contrary to popular belief, the dark night of the soul is not only about pain and misery. It is as much about the freshly dawning light that can break through after the totally debilitating times of the dark night. A helpful book on the subject is Gerald May's *The Dark Night of the Soul: A Psychiatrist Explores the Connection Between Darkness and Spiritual Growth*. Drawing on his clinical experience, May writes that depression and the 'dark night' often go hand in hand, in the same way that we have seen how other spiritual crises can be accompanied by psychotic-type symptoms. He considers it not helpful to try to separate them out, the important thing being to treat the depression where present and to support the 'dark night'.

I have chosen to share with you here Kate's story because she describes so well the different facets of a dark night experience, including the dawning light. Her dark night of the soul spanned, on and off, nearly thirty years. From very early on she was clear that her crippling depression was a spiritual process of purification and cleansing. What she discovered was the more she was able to surrender to it, the easier it became. Here is a condensed version of her story.

"I was pre-disposed to inner worlds from an early age. Most mornings, as a young child, I would wake at five am, open my bedroom window and merge with the energy vibrating from the garden in a state of ecstasy.

Then there was the day, years later, I was cycling through town. As I took in buildings and busyness, I was struck by a profound dissatisfac-

41

tion. My five senses suddenly weren't enough. There had to be more to 'reality' than this seemingly cardboard cut-out three-dimensional existence. It was as if I was trying to connect with something I intuitively sensed and on some level already knew.

My spiritual life started in a conscious sense when aged sixteen, I began having ecstatic mystical experiences along with a dramatic expansion of consciousness. I also experienced the first of a number of episodes of extreme depression — egoic reactions to the pain and negativity my newly burgeoning spiritual life was dredging into consciousness.

Each episode had in common internal dynamics seemingly uninfluenced by whichever anti-depressant I was placed on. The term 'depression' in some ways is apt. It felt like my essence, the very juice of my life force, personality and identity had been pressed out of me, leaving just an empty, shriveled bodily shell. The pain was unspeakable. But as it turned out, the 'bad stuff' — all the anxiety and grief that triggered things in the first place was also being squeezed out. This horrific tight, black space was paradoxically a place of transformation.

I actually felt the moment the shift took place during the fourth episode lasting nine months. As a wave of darkness so black threatened to flush me down the abyss for good, I 'offered up' the experience to something greater than myself.

Like a tiny switch, something flicked inside. From that moment I began infinitesimally to move out and up, like a diver rising from the depths, slowly, slowly so as not to get the bends. It took six weeks to come to the surface. But it didn't stop there. A sense of feeling 'intensely alive' would grow and expand. There wasn't enough time in the day to do everything that inspired and excited me. Underpinning all this, deeper dynamics were also shifting. I literally bore witness as dense blocks of pain, grief and fear transformed into joy, enthusiasm, confidence and purpose.

I felt like an onion being peeled through numerous layers to its core. Each cycle worked on a specific part of my ego. There was something systematic about these virtually symmetrical dynamics within ever deepening cycles. It was as if some kind of super mind had conducted a carefully conceived, sequential process that through each experience purged and prepared me for the deeper and more painful excavation to come.

The fifth and sixth episodes unfolded over the course of five years. The two, 12 month long, 'super-depressions' each alternating with 'intense aliveness' felt like both a bottoming out and culmination of the entire process. I won't go into the horror of depression so extreme I couldn't walk, talk or eat properly for months. The noises that vomited from my mouth were beyond animal. The pain felt primal, as it went back to the dawn of time.

By the time I was surfing my last major 'high' in 2001, the world was online. I also came across the terms 'spiritual emergency' and 'spiritual crisis' during this time and realized this is what I'd been living with on and off for so long. Above all, I discovered I wasn't alone. Thousands of people world-wide from all faiths and none were having similar experiences and the numbers were growing. Some were triggered through 'searching' practices such as yoga, prayer or meditation. Others like myself were undergoing 'spontaneous awakenings'.

The 'super-depressions' were such profound experiences of breakdown and breakthrough that for a while I thought I'd 'made it.' The subsequent years have of course shown yet again that this is one psychospiritual road trip that doesn't end. However, I do have a sense of having broken the backbone of my particular load.

When waves hit, they are smaller ones I can withstand as opposed to the old tsunamis. Needs, desires and expectations have been replaced by more philosophical hopes with a strong dash of 'que sera sera'. I have a sense of deep healing, not just of hurt from this life, but lifetimes before.

I feel clearer and lighter. It's like I've broken through a series of psychological glass ceilings to find a place of freedom and peace that is 'beyond'. That Buddhist thing about non-attachment — I'm starting to live it as opposed to talk about living it. I increasingly feel the paradox of being in the world but not of it."

Kate's is an extraordinary story of guts and determination, of having the courage to trust her intuitive wisdom that the process was one of healing and growth. No matter how horrific it felt. She shows us how the light of dawn can and will follow the dark of night.

We will explore the dark night of the soul more in Chapter 6, when we look at some of the mystics and St. John of the Cross's own dark night.

WHAT TRIGGERS THESE CRISES OF SPIRITUAL AWAKENING?

Often the process of spiritual unfolding happens through our conscious, deliberate spiritual exploration or searching. We start attending classes or workshops, maybe feel drawn to a particular teacher. Or perhaps we begin to introduce spiritual practices, such as prayer, chanting or meditation, into our daily routine. We start developing new friendships with those who share an interest in the path we are exploring.

Sometimes though a major life event, such as bereavement, divorce or an accident, can act as a turning point. From living a seemingly secular, material life, we find ourselves going through a powerful transition, which results in a new world view. We had not intentionally set out on a spiritual journey, but we nevertheless find ourselves on one.

In the same way, people fall into these two different types of triggers for spiritual emergency; those who are actively engaged in spiritual development and those for whom some major life event catapults them into crisis, seemingly out of the blue. So the potential triggers of spiritual crisis fall into these two broad categories. On the one hand, the range of spiritual practices we use in our spiritual development and, on the other, life events and situations, that are often outside our control.

Spiritual Practices

Emma Bragdon, in her excellent book *The Call of Spiritual Emergency*, writes that in our Western religious traditions, the *psychological and physiological* aspects of profound spiritual growth are not widely known. In the Bible, it is not made clear that this is what spiritual awakening involves. Whilst in the texts of Yoga, Hinduism, and Buddhism, it is clearer that there is a psychophysiological result of spiritual practices.[5]

From this angle we can see that practices such as meditation, prayer and ritual can all potentially trigger spiritual crisis, especially if practiced intensively, such as within a retreat context. These practices are designed to help us develop spiritually and to open us up, and that is precisely what they do. Other practices, such as yoga, Chi Gung and tantric sex, which all come from ancient esoteric traditions, can have the same effect. In her book, Emma Bragdon has a whole chapter on sexual experiences and spiritual emergency.

In a study[6], Kaia Nightingale, of the Canadian Spiritual Emergence Service, asked participants what the various factors contributing to their profound experiences were. Nearly two-thirds cited meditation, 29% listed yoga and nearly a third mentioned prayer; 14% named sex.

Many different spiritual development workshops can also be powerful catalysts. Emma, the Shiatsu therapist, sounds a cautionary note.

"To encourage my connection with Chi and improve my Shiatsu, I took up Qigong. I joined a summer ten-day Qigong camp on Dartmoor and that triggered another episode. Gathering in so much Chi all day everyday was too much for me. I became what I've since heard termed a 'Qigong casualty.'"

EMMA

Other workshops and retreats have also been known to spark intense spiritual crisis. Kaia Nightingale found that 35% of participants in her research said workshops were a trigger for them. This is not to say that there is anything 'wrong' with the workshops necessarily, although in some cases it may be that the leaders have no expertise in this area or that there is no system of after-support in place once the workshop is over.

If we follow a Do-it-yourself or mix and match approach to our spiritual exploration, as many do, then we are perhaps in danger of not being fully grounded in any one tradition, of not having the continuity of a teacher who knows us deeply and can guide us when needed. Having said that, even some experienced spiritual teachers can find themselves out of their depth in this territory.

Travel

In the various books on spiritual crisis or emergency, travel is not mentioned as being a potential activator of spiritual crisis. Our experience, however, through the Spiritual Crisis Network, suggests that it can play a large part. Annabel, the massage therapist, tells how her first crisis came about.

"At the age of 28, I made a trip to India where I stayed with a Hindu family for nine months. At the beginning of my stay, I had a powerfully transformative experience. The trigger may have been culture shock. Without any warning, I entered a process which included spontaneous deep breathing and culminated in an ecstatic state. I heard music of a divine and exquisite nature, including angelic choirs and organ music. I have never before or since experienced such a heightened state. With this powerful opening came a plunge into chaos, and I now understand that I had a powerful spiritual experience without a strong enough container to hold it."

My 2003 crisis took place in Egypt. Both Annabel and I visited countries that have considerable spiritual heritages. Our journeys held an element of modern day pilgrimage. This has long been an important spiritual practice and, like those mentioned above, is designed to open us up.

Three other things may contribute to foreign travel sparking spiritual emergency. One is the impact of the vastly differing cultures, as Annabel mentioned. We are catapulted out of our everyday routines into environments where our senses are stimulated at every turn. It is far easier in such circumstances to be totally in the 'now', in the present moment. What happens is that the environment powerfully and automatically creates complete mindfulness in us. We experience the full *Power of Now*, which is profound in terms of spiritual awakening.

The second is that, unlike our ancestors, whose pilgrimage would most likely have been done on foot with a walking stick, we tend to arrive at our destinations by plane. Flying can be extremely ungrounding. One person refuses to fly since her spiritual emergency, the trigger of which she attributes very much to travelling by air.

The third aspect of foreign travel or pilgrimage, which can act as a trigger is that many of the places we visit have a far hotter climate than we are used to. The intense heat and energy of the sun can blast our crown chakra open in no time at all, especially if we are already un-grounded from flying and are feeling the culture shock of the onslaught of unfamiliar colors, noise, smells and more.

PRACTICING SPIRITUALLY OR NOT?

Grof wrote, some twenty years ago, that a deep involvement with spiritual practices, like those we have looked at, was the most important catalyst of spiritual emergency. Whilst engagement in spiritual practices is still a very important trigger, Kaia Nightingale's research in Canada suggests a slightly different picture there today.

She found that as many as 43% of those involved in her study had not previously been practicing in any way. In other words they had not consciously been engaged on a spiritual path. What this possibly suggests is that profound spontaneous spiritual experiences, that appear to come out of the blue, are happening to more people now. This may relate to what I call 'the dark night of the globe'. As we go through transition many of us are experiencing the global awakening of consciousness at an individual level. We are getting a bolt out of the blue to help wake us up.

Both Kimberley and Emma stress that they had not been seeking spiritually before their experiences began.

> "All of this before I had read one book, been to one workshop or heard one teacher speak about such things. I was totally unaware of what any of this meant, which made it all so much more frightening and confusing."
>
> *KIMBERLEY*

So if we are not consciously engaged in spiritual development what sort of things can trigger such an overwhelming experience?

Life events
Anything from a car accident to an operation can trigger spiritual emergency. Kaia Nightingale's study found that illness, burnout and stress were all contributory factors for some people. And as Emma Bragdon points out, stress can include the physical stress the body goes through at times, such as that experienced by athletes or women during childbirth. Janice, who tells her story in Hazel Courteney's book *Divine Intervention II*, was coping with the stress of her husband having had a stroke when she went into crisis. Nightingale also found that 12% cited lack of sleep as one of the triggers. Anything that weakens the body and lowers psychological resistance can precipitate the process.

Loss of any kind, whether through death and bereavement, divorce, loss of a job, home or relationship, can be a catalyst.

> "My experience began in the February of 1998, two months after my mother died."
>
> *KIMBERLEY*

Loss of faith, which can be utterly devastating, is specifically mentioned under the category Religious or Spiritual Problem in the DSM, the diagnostic manual used by American psychiatrists.

First Love
Remember falling in love for the first time? For some, in their teens, at a delicate time of psychological development, the rawness and intensity of falling in love for the first time can trigger the beginnings of spiritual emergence, or even full-blown crisis. Kate's mystical experiences first began then and Aaron's spiritual emergency was also partly triggered by his first love.

Timing
Easter is a particularly potent time of year in the West and a number of people seem to go through spiritual crisis around then. Hazel Courteney writes about her experience and that of others happening at Easter. With the Christian focus on the crucifixion and resurrection it is a powerful time of death and rebirth, such a central issue in spiritual crisis. It is also a time when the theme of good versus evil, so prominent in many spiritual emergency experiences, can come to the fore. The archetypal energies of Jesus

Christ and Judas Iscariot are once more with us. It is also springtime, with new birth, new growth.

Other times of the year that can be powerful are equinoxes, solstices and full moons, in particular the full moon in May. Millions of Buddhists around the world celebrate this Wesak full moon as being when the Buddha gained enlightenment.

Drugs

We can distinguish between three different types of drugs: street drugs; drugs prescribed for medical reasons; and anesthetics. This was one person's experience:

> "I went to University and smoked lots of dope. So my story begins with the dope smoking, in December 1996. I was trying to give it up, after about 6 years of daily smoking. The problem was that if I didn't have my usual joint before bedtime I couldn't get to sleep. I didn't know how to relax without it. The opportunity came to give it up when I discovered that the man I was seeing was also seeing someone else. I ended up smoking a large bag of skunk and then promising myself that I would never buy any more. I was determined to give it up and now was the time. As expected, I didn't sleep. I figured that I would eventually fall asleep, but after five nights I became worried that I might actually die. My doctor was no help at all and told me that I was sleeping, I just didn't realize it. He also prescribed tranquillizers rather than sleeping tablets and these didn't get me to sleep either. So a friend took me back up North to my parent's house and by this point I couldn't really communicate properly. I was taken to hospital in an ambulance."

Each of the three different types of drugs has the potential to trigger spiritual crisis and although it is less common with medical drugs and anesthetics, Grof certainly came across examples where a drug prescribed for medical reasons triggered the crisis. Emma Bragdon, in her book, shares the story of a woman whose crisis was triggered by the anesthetic used during a visit to the dentist.

Life Stages

Bragdon sensitively explores the issues that can give rise to spiritual crisis during the different phases of our lives, right through to preparing for death. She talks about the psychological work that needs to be completed at each developmental stage and the problems that can be encountered if that does not happen, including in childhood and adolescence. These are rarely mentioned elsewhere, almost as if spiritual emergency were the exclusive domain of adults, which of course it is not. Many youngsters going through the transition to adulthood experience spiritual crisis. Aaron's tragic story shows the very real dangers at this stage in a young person's life.

Many young adults in their early twenties, who find themselves going through spiritual emergency, are in effect still grappling with the developmental issues of adolescence; of establishing an identity in the world and separating themselves from the family. Kimberley was 25 when her crisis began.

Many Triggers and None

We have seen that a whole range of things can act as the catalyst for spiritual emergency. Sometimes it can be difficult to pinpoint what caused it. There can be no obvious or apparent trigger whatsoever. At other times it can be a combination of several things.

> "Inspired by my Indian trip I began to meditate daily on my return to the UK, sometimes for up to an hour. I also had a job as a nurse in a hospice. The trigger on this occasion may have been a combination of stress, the springtime of the year, and the meditation practice, alongside susceptibility due to the lack of rootedness I had in my own body."
>
> *ANNABEL*

The long list above of what can trigger spiritual crisis shows that virtually anything can, if the circumstances come together. This suggests, says Grof, that the person's readiness for inner transformation is far more important than external triggers.

In this chapter we have explored the key features of spiritual emergency and the sorts of things that can potentially trigger it. We have glimpsed the intensity of such experiences through the eyes of those who have been through it. Let us take a look now at what the scientific and medical community has to say about these extraordinary processes of psychospiritual transformation.

CHAPTER 3

The Medical and Scientific Background

Today there is a whole body of contemporary research and literature supporting our understanding of spiritual emergency. Much of it comes from within a growing branch of psychology known as transpersonal psychology, 'trans' meaning 'beyond' the personal; in other words a school of psychology that encompasses the transcendent, the spiritual dimension. It brings together ancient mystical wisdom with modern psychology, grounded in scientific research.

Another branch of psychology, which overlaps and where much quantifiable research has been carried out, is that of parapsychology—the investigation of the paranormal. This field sheds much light on the kind of anomalous or unusual experiences that those going through spiritual emergency often have. When mental health professionals are open to such findings, it can help put into context 'symptoms' that might otherwise be pathologized as illness. Here we are talking about a whole range of psychic abilities, such as precognition or extrasensory perception that those going through spiritual crisis often report. A full survey of parapsychology is beyond the scope of this chapter, but you might like to look at the work of Dean Radin [7] or Charles Tart [8].

THE TRANSPERSONAL LINEAGE

Transpersonal psychology grew out of humanistic psychology, the main proponent of which was Maslow, best known for his theories on the hierarchy of needs, peak, or mystical, ecstatic experiences and self-actualization. We can trace a distinguished transpersonal lineage down from Maslow through the likes of Assagioli, Jung, Perry and Laing, down to Grof, the psychiatrist who, along with his wife Christina, is credited with coining the term 'spiritual emergency'.

When I first read Stanislav Grof's work I had an 'aha' moment. With enormous relief, so much fell into place for me. Here was somebody who was offering a framework that made sense, with which I could interpret my very strange experiences. Here was somebody, a psychiatrist, a respected member of the medical profession, explaining in language I could understand, that what I had been through was a recognized phenomenon. And I was not alone. One colleague from the Spiritual Crisis Network told me she cried tears of relief when she discovered Grof.

Before we look at Grof's contribution today, let us go right back to the beginning of the 20th Century, to the first Western attempts to scientifically map the psychology of spiritual experience.

BUCKE

1901 was the year that Queen Victoria died and that Teddy Roosevelt became President of the United States of America. It was also the year that a retrospective Van Gogh exhibition in Paris caused a sensation and the year that Bucke's *Cosmic Consciousness* was published. Richard Bucke, a Canadian psychiatrist, investigated the state he called Cosmic Consciousness and identified certain key features of it, based in part on his personal experience:

> … a consciousness of … the life and order of the universe … an intellectual enlightenment or illumination … a state of moral exaltation, an indescribable feeling of elevation, elation and joyousness, and a quickening of the moral sense … a sense of immortality, a consciousness of eternal life …[9]

Bucke saw cosmic consciousness as a perfectly natural and normal progression, 'a matter of course — an inevitable sequel' of man having become self-conscious:

> … our descendants will sooner or later reach, as a race, the condition of cosmic consciousness, just as, long ago, our ancestors passed from simple to self consciousness.[10]

Cosmic Consciousness is a fascinating historical document. It is also very much a product of its male-dominated, patriarchal era. Of those he found to have been in a state of cosmic consciousness, Bucke includes the Buddha, Jesus Christ, St. Paul and Mohammed. He also covers Dante, Francis Bacon, William Blake, Balzac and Walt Whitman, amongst others. There is no mention of any women, such as Hildegard of Bingen or Julian of Norwich, to name just two of the many he could have included.

JAMES

If Bucke's work was the first 20th Century attempt in the West to research and chart spiritual experience, James's *The Varieties of Religious Experience* followed hot on its heels. Published the very next year, in 1902, the text was the result of a series of lectures the American William James was invited to give at the University of Edinburgh, Scotland. James focused on the psychology of personal spiritual experience at a time when psychology was in its infancy.

In the same way that researchers later experimented with LSD in the 1960s, he used the equivalent of his time, in his efforts to understand the mind and spiritual experience. The substances he experimented with included chloral hydrate, one of the recreational drugs of the day, nitrous oxide, also known as laughing gas, and peyote, the plant used for sacred rituals by indigenous Americans. Holding a series of academic posts at Harvard University from 1873, in the psychology and philosophy departments, he taught, amongst others, the young Roosevelt, before he went on to become President.

James emphasized above all the importance of personal spiritual experience, partly because of what he referred to as the 'ineffability' of mystical experience; that such experience cannot be communicated, it defies words or expression, it is beyond the rational mind: "It follows from this that its quality must be directly experienced; it cannot be imparted or transferred to others." [11]

Aspects of mystical experience can feature prominently during spiritual emergency and James was the first to study the psychology of mystical states. We can follow the transpersonal lineage on from him.

ASSAGIOLI

In 1910, eight years after James's *The Varieties of Religious Experience* was published, the Italian Roberto Assagioli completed his doctoral research on Psychosynthesis. This is the name he gave to the transpersonal, or spiritual, psychotherapy that he founded.

Assagioli contributed enormously to our current understanding of spiritual emergence and emergency. His paper *Self-Realization and Psychological Disturbances*, originally published in English in 1937 is reproduced in Grof's edited volume *Spiritual Emergency: When Personal Transformation Becomes a Crisis*. Assagioli made a distinction between self-realization and Self-realization:

> The meaning most frequently given to self-realization is that of psychological growth and maturation, of the awakening and manifestation of latent potentialities of the human being… These correspond to the

characteristics Maslow ascribes to self-actualization ... full Self-realization [is] ...where the personal-I awareness blends into awareness of the spiritual Self. ...It is not the realization of the personal conscious self or 'I', which should be considered merely as the reflection of the spiritual Self... [12]

Assagioli saw the critical phase of spiritual crisis as having four distinct stages:

• preceding the awakening
• the awakening itself
• reactions to the awakening
• the process of regeneration of the personality in the service of Self-realization.

Today Psychosynthesis is taught around the world. In London alone there are three Psychosynthesis schools. Trainee psychotherapists cover spiritual emergence and emergency in their studies.

JUNG

12 years after James traveled from America to Scotland to give his lectures on *The Varieties of Religious Experience*, Jung too would travel to Scotland to give a lecture. It was the summer of 1914; World War I was about to break out. At this point Jung had already started going through his own process of spiritual emergence and emergency, which we will explore fully in Chapter 6. He had seen horrific visions of Europe flooded in blood, which, as a psychiatrist, he thought presaged his imminent mental breakdown. As he prepared his paper on schizophrenia he repeatedly said to himself: "I'll be speaking of myself! Very likely I'll go mad after reading out this paper."[13]

With the outbreak of war, Jung realized that in his horrendous visions he had been picking up on events to come, in a precognitive way. He sought to understand how that could be, how that could happen. Starting from his understanding of the personal unconscious, he came to see that there is a whole other realm that can erupt from the unconscious that is far broader, far more universal. He called this the collective unconscious.

He had also experienced first-hand, with those visions, how easy it is for material from the personal and from the collective unconscious to get mixed up, how difficult it is to separate them out. When the volcano of the unconscious erupts, the lava and ash of the personal and collective unconscious are all spewed to the surface together.

Jung's theory on the collective unconscious is closely linked to what he called 'archetypes', which are common to us all, across time and cultures. Before using the term 'archetype' Jung referred to these as 'primordial images' or 'archaic remnants'. These motifs or images are found in mythology and access to them lies deep within our psyches, inherited from distant times and civilizations.

Jung's understanding of archetypes first came to him at the age of 35, when he read about soul-stones and *tjuringas* or *churingas*. These are sacred stone or wooden objects, used in rituals by indigenous Australians. This triggered a childhood memory of having hidden a small figure, along with a small oblong stone, in a box in the rafters of the attic. It was his boyhood secret that no one knew about. In his reading he discovered that he had been acting out an ancient religious rite. There was no way he could have consciously known about such rites; there was no book in his father's library that covered such topics[14]. This was the beginning of his understanding of archetypes and the way in which they are universally available to us through the collective unconscious.

We probably all have examples of these forces at work in our lives. As a teenager I mysteriously found myself in art class painting large, white, winged horses flying over a mid-night seascape bathed in the light of the full moon. I had no idea where the image had come from, nor did I know that I was painting Pegasus, the divine horse that enables us to access the realm of the Gods, to transcend from one plane of consciousness to another.

Jung's ideas about the personal and collective unconscious and archetypes are extremely important for anyone going through spiritual crisis. These energies inhabit the deepest recesses of our psyches. As we shift into an altered or non-ordinary state of consciousness, the unconscious, both personal and collective, and its associated archetypal realm, erupt to the surface, wreaking havoc.

Archetypal events, such as birth, separation from parents, marriage, death, the union of opposites, will feature strongly in our concerns. Archetypal motifs, such as the Apocalypse or Creation, which appear in our dreams or even as visions in our waking state during spiritual crisis, can be at best confusing and at worst terrifying if we do not have an understanding of their nature. Archetypal figures too, such as the Goddess or the Devil are more than likely to make an appearance during spiritual emergency.

As well as his theories on the collective unconscious and related archetypes, Jung is also known for his ideas on the psychology of spiritual experience and the process of individuation. Jung said he saw individuation, the process of 'coming to self-hood or self-realization' as 'the central concept of my psychology'. In as much as spiritual emergency is the psyche's attempt to evolve towards self-realization it can be equated with Jung's concept of the individuation process.

Jung was ahead of his time. That he was able to be such a pioneer, leave such a legacy and ultimately make such a contribution to the field of transpersonal psychology is, I believe, largely thanks to his own personal experience of spiritual emergence and emergency. We will look at this in more detail later.

LAING AND PERRY

Following Jung in the transpersonal lineage come the British psychiatrist and psychotherapist Ronnie Laing and his American counterpart John Perry. Both stand out for their radicalism. Both challenged conventional psychiatry and both were ultimately rejected by the mainstream for it. This was not before they were able to secure funding, in the liberal atmosphere of the 1960s, and carry out research into alternative residential settings for in-patients.

Laing set up the Philadelphia Association in 1965, with Kingsley Hall in London being the first project, whilst Perry set up Diabasis in San Francisco a decade later. Perhaps with the advantage of learning from Laing's earlier venture, Perry seems to have been more cautious with admissions, taking mainly only those who were presenting with their first psychotic break.

His approach included using little or no medication because he found the way it suppressed the deep contents of the psyche to not be helpful. In his book *Trials of the Visionary Mind: Spiritual Emergency and the Renewal Process* he reported the results of outcome studies and a three-year follow up. These showed patients treated without medication faring markedly better than those treated with, both in terms of recurrence of 'psychosis' and emotional development. But as he says:

> The reports of those findings have been ignored almost totally, perhaps because this kind of news is unwelcome and inconvenient.[15]

Meanwhile the work of Laing's Philadelphia Association continues to this day. Since its controversial beginnings, the Association has run over 20 community houses, today more as therapeutic communities than alternatives to acute in-patient settings.

When I first read both Laing and Perry, I was fascinated by their findings; so much of it made sense in light of my own experiences. Laing saw a psychotic breakdown as "a healing process that if allowed to occur naturally will result in the reintegration of the personality". Laing's research into schizophrenia led him "to consider the similarity between certain kinds of apparent madness and valid spiritual experiences".[16] As part of his own rich spiritual life, Laing spent a year in Sri Lanka, studying Buddhism and meditation. The following is from his contribution to Grof's volume *Spiritual Emer-*

gency: When Personal Transformation Becomes a Crisis:

> True sanity entails in one way or another the dissolution of the nor-
> mal ego … the emergence of the 'inner' archetypal mediators of divine
> power, and through this death a rebirth, and the eventual reestablish-
> ment of a new kind of ego-functioning, the ego now being the servant
> of the Divine, no longer its betrayer.[17]

Laing and Perry used slightly different language. Laing did not use the term 'spiritual emergency', but both saw 'psychosis' as potentially healing and regenerative. Perry
focused on the process of spiritual emergency as a natural process of renewal, of the
psyche seeking to heal itself and move towards a higher level of functioning, a higher
level of awareness and consciousness. His clinical practice led him to conclude that
very often a person's first acute 'psychotic' episode was:

> nature's way of healing a restricted emotional development and of lib-
> erating certain vitally needed functions — in short, a spiritual awaken-
> ing.[18]

Where mainstream psychiatry refuses to listen to or value patients' non-rational talk,
he, like Laing, listened for the psyche's deeper concerns. He understood the importance of "the unconscious's non-rational system of meaning and values", much as
Jung before him had rated its apparently irrational content. "The disadvantage of the
non-rational systems is that they are not granted recognition and validation by our
culture…"[19]

Perry's basic premise is that humans have an inner spiritual life with deep processes
of its own. The psyche needs to be allowed to go through disorganization on the way to
reorganization. He found that patients required a minimum of medication so as not to
interfere with what nature is engaged in doing, so as not to interfere with this natural
'process of renewal.

The phase of disorganization and disintegration needs to be validated rather than
pathologized. When our experience is pathologized we are told that there is something wrong with us, that we are ill, that we need 'fixing'. It is essentially a negative
message. The opposite of that is when our experience is validated by those around us.
The message is very, very different, that what we are going through is okay, that many
other people have also been through it and come out the other side.

This fundamental importance of such validation when going through spiritual
emergency was also emphasized by Assagioli.

> … the therapist can give much help by assuring him [the client] that
> his present condition is temporary and not in any sense permanent or
> hopeless as he seems compelled to believe. The therapist should in-
> sistently declare that the rewarding outcome of the crisis justifies the
> anguish — however intense — he is experiencing. Much relief and
> encouragement can be afforded him by quoting examples of those who
> have been in a similar plight and have come out of it. [20]

The alternative crisis home, Diabasis, which Perry set up, was distinct from usual
in-patient settings. The home had facilities such as an art room and a 'rage room',
equipped with mattresses where people could thrash about safely and vent their feelings. This is precisely the kind of room that Aaron, the teenager who died tragically,
instinctively knew he needed in order to be able to safely express his emotions.

One of the main distinguishing features of Diabasis was that many of the staff had
personal experience of extreme or non-ordinary states of consciousness, rather than
necessarily being traditionally trained mental health staff. The advantage of such per-
sonnel is that they will both value the extreme and challenging aspect of the person's
experience and at the same time, vitally, will not be afraid of it.

> … what is absolutely crucial to the individual who is beginning to 'space
> out' is the emotional response of the persons in his surroundings. If they
> are appalled, bewildered, and afraid … then the individual in the plight
> feels himself isolated and alone. Panic seizes him…[21]

What Perry found at Diabasis was that if you validate the person's experience and support
it in the right way, giving people a positive message about what they are going through,
even very 'psychotic' clients will "usually come into a coherent and reality-oriented state
spontaneously within two to six days".[22] Moving through such extreme states of consciousness or spiritual emergencies to a place of integration then took six to eight weeks
on average. In Perry's book *Trials of the Visionary Mind: Spiritual Emergency and the
Renewal Process* you can find a research report on Diabasis (Appendix C).

This vitally important difference in clinical outcomes between validating and
pathologizing a person's experience has also been identified by the user-led Recovery
Movement. This is where the transpersonal field and the Recovery Movement overlap,
in that research has shown the protective qualities of spirituality, in preventing relapse
and in finding a meaning and sense of purpose in life. Key figures in the Recovery
Movement have been Mary Ellen Copeland in the USA and the psychiatrist Glenn
Roberts in the UK.

GROF

Whilst Laing and Perry were researching alternatives to psychiatric in-patient settings, Stanislav Grof was researching with LSD. The 1960s were in full swing. Having completed his medical and psychiatric studies in his native Czechoslovakia, Grof first conducted LSD research at the Psychiatric Research Institute in Prague and then at the John Hopkins University School of Medicine in Baltimore, USA. He went on to become Chief of Psychiatric Research at the Maryland Psychiatric Research Center.

Grof has spent 50 years researching non-ordinary states of consciousness. When LSD became illegal he discovered that similar states could be explored using a particular breathing technique. He called this 'Holotropic Breathwork', referring to the healing movement towards wholeness of such non-ordinary states of consciousness as meditative, mystical or psychedelic experiences.

In his work on spiritual emergency, Grof has been outspoken about psychiatry's tendency to pathologize non-ordinary states of consciousness. Of his many books, two in particular deal with spiritual emergency; *Spiritual Emergency: When Personal Transformation Becomes a Crisis* and *The Stormy Search for the Self: A Guide to Personal Growth through Transformational Crisis*. Both are co-authored with his wife. The first is an edited collection with contributions from, amongst others, Laing, Assagioli and Perry.

Whilst these texts are over 20 years old now, they still form, along with Bragdon's *The Call of Spiritual Emergency*, an essential starting point for anyone new to the territory.

All of these transpersonal clinicians and researchers point to one aspect of the experience of spiritual emergence and emergency. Whether it is Maslow's peak experiences, Perry's take on spiritual crisis as a renewal process, or Jung's process of individuation, all their work points to one thing; to spiritual emergency as a natural, healing, indeed healthy, process. For me, reading their work confirmed my innate sense of it as a positive process, one, nevertheless, which needs careful guidance and considerable support in order to fulfill its promise.

CHAPTER 4

A Survey of
Relevant Research

In their research the Grofs identified a typology, a classification of the different types of spiritual emergency they found. In the 20 plus years since then much research has been carried out into these different dimensions of spiritual crisis. This has served to strengthen the scientific validity of the field. Without going into the academic research in detail, my aim in this chapter is to point towards this body of evidence. If you are interested, you can follow up some of the references and suggestions for Further Reading. The main types of spiritual emergency the Grofs identified are:

- Peak experiences
- Kundalini awakening
- Near-death experiences
- Past-life memories
- Psychological renewal
- Shamanic crisis
- Psychic opening
- Communication with spirit guides and channeling
- Close encounters with UFOs
- Possession or obsession states

These can most usefully be seen as the different elements that may combine in any one individual experience of spiritual emergency. In other words, elements from the different categories are likely to overlap in a person's experience. Let us take a look at some of these aspects of spiritual emergency that have been more thoroughly researched.

KUNDALINI AWAKENING

When we looked at some of the key features of spiritual emergency, we touched on the process of Kundalini awakening. Some of the elements described, such as unexplained spasms and jerking, feelings of vibration in the body, burning sensations of searing heat in parts of the body, or all over, are all typical of Kundalini awakening.

This life-force energy can be activated by spiritual practices, such as Kundalini yoga which has become so popular recently. It can also come to life spontaneously, seemingly from nowhere. This is what Olga Louchakova, director of the Neurophenomenology Center at the Institute of Transpersonal Psychology (USA) found:

> I first recognized that the Kundalini process can include physical problems in the late 1980's. I had been working as a medical doctor and neuroscientist while also teaching intense Kundalini-raising practices in the Russian underground community during the years of Soviet power. … As my observations cumulated over the years, it became clear to me that health problems can develop in people who do not practice Kundalini techniques but are the subjects of spontaneous Kundalini awakening. [23]

Difficulties that can emerge during spiritual emergency related to Kundalini awakening are particularly difficult to identify because they can take so many different forms. As such symptoms are often mistaken for physical or psychological issues that fit into familiar Western medical diagnostic categories. As Bruce Greyson, Professor of Psychiatry and Neurobehavioral Sciences at the University of Virginia, tells us:

> In Eastern traditions, Kundalini would ideally be activated at the appropriate time by a guru who can properly guide the development of that energy. If awakened without proper guidance … Kundalini can be raw, destructive power loosed on the individual's body and psyche. [24]

Gopi Krishna [25] wrote extensively about this raw power, based on his personal experience, and really put Kundalini on the map. The American psychiatrist Lee Sannella [26] then took up the baton and researched the phenomenon extensively, building on Bentov's [27] model of physio-Kundalini. He identified four key features: motor, sensory, interpretive and neurophysiological phenomena (which equate to the symptoms mentioned earlier). His findings were further corroborated by the work of Bonnie Greenwell. [28]

Taking Sannella and Greenwell's work further, there are several Kundalini research centers in America and the Kundalini Research Network has been active in the USA since 1990. Their conferences boast an impressive line-up of speakers from the medical community. Fields represented in a recent text, *Kundalini Rising* (Sounds True 2009),[29] include neurophysiology, the branch of physiology that deals with the functions of the nervous system, and neurophenomenology, which combines neuroscience with phenomenology in order to study experience, mind, and consciousness.

Whilst current Western understanding and research into Kundalini has come from Indian yogic teachings, equivalent energies are known in many other cultures, from Tibetan yogis and Eskimo shamans to Christian mystics.[30] As Kundalini awakening becomes more thoroughly studied and more widely recognized the most pertinent question at this time of global transformation seems to me to be David Lukoff's: "Are the medical and Kundalini support communities equipped to handle the increasing occurrences of spiritual awakenings?"[31]

NEAR-DEATH EXPERIENCES

Near-death experiences and related out-of-body experiences have attracted much attention in recent years and have been the subject of considerable research. These experiences can, as we have seen, trigger spiritual crisis. This was the case for Jung who went through a minor crisis after the heart attack he had at the age of 69. His near-death and out-of-body experiences are described in his autobiography *Dreams, Memories, Reflections*, along with his subsequent difficulty in adjusting back to everyday, mundane life.

As well as many personal accounts that are available,[32] individuals like Robert Monroe researched and reported on the thousands of out-of-body experiences he had over decades in an influential trilogy of books.[33] Important research has also been carried out by people such as Elisabeth Kübler-Ross,[34] Raymond Moody[35] and Bruce Greyson[36] in the USA and by Peter Fennwick[37] and Sam Parnia[38] in the UK. Useful texts include *The Near-Death Experience: A Reader* edited by Lee Bailey and Jenny Yates.[39]

In his research Moody identified key features of a near-death experience (NDE), including a sense of peace and loss of physical pain, if the person had been in pain before, a sense of floating above their own body, of being drawn into a tunnel or rising rapidly upwards, above the earth, meeting relatives, friends or a being of light and love, being shown a life review and feeling reluctant to return.

In terms of our focus on spiritual emergency, the most exciting aspect of this research for me is that it is helping to broaden out consensual reality. The mainstream is increasingly being obliged to consider as legitimate phenomena that were previously dismissed. NDEs and out-of-body experiences (OBEs) are closely related to this next topic, that of past-life memories.

PAST-LIFE MEMORIES

Past-life regression also comes within the field of transpersonal psychology. Again, the research here is mainly qualitative, based on case studies. For millions around the world reincarnation is a given, as much a part of a person's frame of reference as the changing seasons. Indeed it would also have been for those of us brought up in the Christian West, if it had not been written out of the history books by Emperor Justinian (483-565 AD) at the Fifth Ecumenical Church Council in 553 AD.

Emma Bragdon gives a full account of the political reasons for erasing the doctrine of reincarnation from the Christian Church in her book *Kardec's Spiritism*. She points out there are sections of the New Testament that can only be understood in the light of reincarnation.[40]

For those who are not familiar with past-life work, perhaps the most accessible point of entry is the field of trauma studies. Research by Peter Levine,[41] one of the prominent figures in the discipline, shows that trauma can get held or frozen in the body. So, for example, a study of children trapped underground in a landslide showed that those who were able to actively attempt to dig their way out were less traumatized than those who were frozen, disabled by their shock and fear. Any trauma held in the body then needs releasing for full healing to take place.

The same principle applies to past-life trauma. In a healing session the client is helped to release the stuck trauma, by reliving the past-life event. It can seem something of a mystery as to how this body, in this lifetime, a completely different body from the past-life in question, could still be holding trauma and could experience relief from the release of that trauma. Again we need to draw on Eastern teachings to be able to make sense of this. Hinduism and Buddhism refer to *samskaras* as blueprints that are held at a subtle, energetic level, where all our past conditioning and patterns are stored, including any trauma. Roger Woolger's excellent book *Other Lives, Other Selves* explains this very well, including how *samskaras* and trauma accumulate in layers. An example might be someone who was born with the umbilical cord wrapped around their neck, later in life suffering a neck injury. A past-life session might reveal that they were hanged in a previous lifetime. I cannot help thinking that *samskaras* were at play when world-famous dancer Isadora Duncan died tragically in a freak automobile accident in 1927. Her flowing silk scarf caught in the open-spoked wheel of the sportscar she was travelling in, strangling her to death and nearly decapitating her in the process.

My own process of spiritual emergence and emergency included having to work through some horrific past-life material. Unfortunately, the happy, easy, trauma-free lives tend to not come up to the surface, as they do not need healing. We can find ourselves reliving extremely disturbing scenes. What happened for me, and this can be fairly typical, is that only fragments started to come to the surface, because that was all I could cope with initially, all I was ready for.

Once I realized what was happening, that the images and snatches of scenes I was seeing in my mind's eye were past-life related, I sought professional help. A chapter in Roger Woolger's book, *Eros Abused: the past life roots of sexual and reproductive problems* spoke to me so strongly that I sought out his workshops and did some training with him. What particularly appealed to me about his regression work is that he does not use hypnosis, which worked very well for me. It is surprising how easy people find it to access past-life material that needs attention. Exploring out of sheer curiosity, or for the hell of it, is probably not such a good idea. If an issue requires healing and is calling to you, you will know about it.

In the sessions I had to deal with layers of appalling sexual violation accumulating over several lifetimes. In one lifetime, living as a young nun in a convent, I was raped and killed by a soldier. These are not easy traumas to relive and release. One manifestation of this particularly challenging sexual *samskara* was in this lifetime, when my first boyfriend was sexually abusive towards me. Patterns can keep repeating themselves endlessly until we are able to bring them into our conscious awareness.

The healing work I did with Roger Woolger transformed my life. Sexual fears and lack of pleasure dissolved, enabling me to go on to form my first ever sexually satisfying, loving relationship based on trust, respect and healthy boundaries. I gained far more than that from the work, however. As I experienced myself as the dead nun, in the after-life state, the *bardo* as it is known in some traditions, my understanding of myself rapidly expanded. Now I can know myself to be far more than just this physical body called Catherine.

This inter-life, *bardo*, state has been the subject of more and more research in recent years. People such as Michael Newton [42] in the US and Andy Tomlinson [43] in the UK have contributed to a deepening understanding of what happens after we leave one incarnation and before we come into the next. Whereas the research of people like Ian Stevenson [44] and Jim Tucker [45] has focused more on children's memories of past-lives. Their research papers and books are brimming with fascinating accounts, accounts where it has been possible to verify facts reported by children about their previous lives.

Not everybody experiences past-life material coming up during spiritual emergency, but it is worth being alert to the possibility, so that appropriate help can be sought if needed.

POSSESSION OR OBSESSION STATES

In their findings the Grofs referred to 'possession states', a term which can be highly emotive and is liable to be misconstrued. The word 'obsession' seems, however, to more accurately reflect the way in which spirit entities can enter our energy fields, persistently causing a whole range of problems. Full possession states are extremely rare and the somewhat loaded term conjures up unfortunate connotations of X-rated movies and exorcisms.

Also known as spirit attachment, here is the Spirit Release Foundation's explanation of the phenomenon, taken from their literature:

> A minority of those who die fail to make the transition successfully. They become what is known as earthbound, because they remain mentally attached to the earth plane and so cannot progress. Reasons for this include a traumatic death, concern over some unfinished business or anxiety for a loved one on Earth.
>
> Such earthbound spirits are rarely malicious: mostly they are lost and confused. This may manifest in a variety of ways. They may attach to a house or locale with which they were associated in life, so that the house or locale becomes haunted. Such spirits may also be responsible for poltergeist activity, which is their way of drawing attention to their plight.
>
> In some cases an earthbound spirit may attach to a living person. Some of the more common symptoms can be: lack of energy, memory disturbance, behavioural change, mood change, addictive behaviour, relationship problems and hearing disturbing voices. There may be bodily pain and other physical symptoms. The degree of attachment also varies. Some individuals are scarcely affected, while in rare cases the individual's body and mind have been taken over completely.[46]

In the UK, the Spirit Release Foundation, set up by psychiatrist Alan Sanderson, is active in working with clients to help attached spirits move on. The Foundation offers accreditation courses for spirit release practitioners. Equivalent courses are offered by the Rev. Judith Baldwin in the USA.

When it comes to understanding the spirit realm and its relationship to mental health struggles, the Brazilians are streets ahead of us. Why is this? Culturally, the Brazilians do not have a problem with acknowledging the realm of spirit. As a nation they are racially very integrated, as Emma Bragdon explains in *Kardec's Spiritism*.

Generations of indigenous Brazilian Indians, European colonizers and Africans who were taken there as slaves have married each other. In the process their Shamanic and Christian faiths have merged. The result of weaving these threads together is the approach called Spiritism. It brings together elements from esoteric traditions, such as spirit communication through mediums; Christian beliefs, Indian philosophy on reincarnation and karma, along with the modern science of parapsychology, metaphysics, the new physics and the study of energy healing.

Although there are many strands of spiritist practice in Brazil, it is the school based on the teachings of Frenchman Allan Kardec that has given rise to over 6,500 Spiritist Centers throughout the country. A leading figure and international spokesperson is Divaldo Franco, who speaks at many events, including occasionally in the US and UK.

When I first felt drawn to Brazil, to visit John of God's Spiritist healing center in Abadiania, *La Casa de Dom Inacio,* I was intrigued to discover that Emma Bragdon was one of the official guides. I consider her book *The Call of Spiritual Emergency* to be an absolute classic on the subject. So I was interested to see that her journey had taken her from exploring spiritual emergency to this Brazilian sanctuary of physical, mental and spiritual healing. It turned out to be an entirely logical progression.

When I visited *La Casa de Dom Inacio,* named after Ignatius Loyola who inspired Kardec's Spiritism, I found that Bragdon had been writing about Spiritism. By one of those quirks of international travel, my bank card would not work. I had virtually no cash nor access to any. A total stranger very kindly and trustingly leant me some and the one thing I bought was Emma Bragdon's book *Kardec's Spiritism: A Home for Healing and Spiritual Evolution.*[47] This was my introduction to a Spiritist approach to relieving mental distress. In the book Bragdon tells about a Spiritist Center in Palmelo, central Brazil, which includes an extraordinary psychiatric hospital. The hospital is run on Spiritist principles, combining treatments such as energy passes by healer mediums to help with spirit release, with Western psychiatric approaches, such as occupational therapy and medication. The results are impressive. In 1997 the Center was recognized by UNESCO (United Nations Educational, Scientific and Cultural Organization) as an asset serving all of humanity.

One of the case studies Bragdon tells about in her book involves Marcel, a young man who was diagnosed with schizophrenia. He experienced crabs and spiders crawling all over him and mood swings, including at times turning aggressive. He was told he would have to manage his symptoms with strong medication for the rest of his life, despite the medication making him feel unwell.

Marcel then spent about three months at the Spiritist psychiatric hospital in Palmelo, where he received regular energy healing alongside mainstream treatment. It became apparent to the mediums treating him that he was being troubled by spirit entities, who had known him in past lives. The energy passes he received helped to clear the spirit attachments and he went on to make a full recovery.

For solid research in this field of spirit attachment and release as well as examples of psychiatric good practice we need look no further than Brazil. There is a compelling and convincing body of research in Brazilian academic and medical circles. Papers presented by nationals at the Second British Congress on Medicine and Spirituality (2009) [48] included the *Neuroscience of Mediumship* and *Depression, Bipolar Disorder and Spiritual Disturbances*. Medical schools in Brazil have courses on Medicine and Spirituality. If you are studying medicine or psychiatry and get a chance to do a placement abroad, Brazil could make for a mind-opening destination.

The disturbances caused by spirit attachment can be an issue in mental health generally and more specifically in spiritual emergency. Whilst we are a long way from mainstream services taking this on board, there are tentative signs that some are beginning to pay attention. John Nelson, the American psychiatrist, suggests in his textbook *Healing the Split*, [49] that nonphysical entities taking up residence in a person's psyche could explain multiple personality disorder and the UK Royal College of Psychiatrists' Spirituality and Psychiatry Special Interest Group have run a day-long programme on 'Minds within Minds: the case for Spirit Release Therapy'.

We have been looking at some of the different types of spiritual emergency, or, more precisely, some of the varying elements that can come together in any one person's experience. It is very encouraging that serious academic research has been conducted in areas such as Kundalini awakening, near-death experiences, past-lives and spirit attachment. It is also very useful for those of us going through spiritual emergency. For those who experience it seemingly out of the blue, who have no previous familiarity with such areas, it can be very reassuring. It can provide a framework to understand what we have been through. It can also strengthen our hand enormously when dealing with health professionals or other agencies. It helps establish the credibility and validity of the phenomenon in their eyes. Articles in reputable, peer-reviewed, scientific journals are more difficult to ignore. All this contributes to broadening the consensual reality around spiritual emergency, to pushing against the boundaries of mainstream scientific materialism.

THE TRANSPERSONAL IN THE 21ST CENTURY

In the last chapter we looked at the pioneers of transpersonal psychology, people who helped establish it as a discipline in its own right. Who and what is carrying this specific work on spiritual emergency forward into the 21st Century?

DSM

Developments in the transpersonal or spiritual approach to mental health have mirrored each other across the Atlantic. In 1999, the Spirituality and Psychiatry Special Interest Group was established in the Royal College of Psychiatrists, UK, under the guidance of its founder chair, Dr. Andrew Powell. This has a large and growing membership of over 2,600 psychiatrists, which represents about one sixth of the College's membership, with an active program of seminars and frequent publications, such as conference papers and books.[50]

Only five years before, in 1994, a new category in the DSM, the Diagnostic and Statistical Manual of the American Psychiatric Association, was introduced. The category 'Religious or Spiritual Problem' was the first time it had been recognized that, where there was a strong religious or spiritual aspect to a person's mental health problems, it was not necessarily a sign or symptom of illness per se. It states:

> This category can be used when the focus of clinical attention is a religious or spiritual problem. Examples include distressing experiences that involve loss or questioning of faith, problems associated with conversion to a new faith, or questioning of other spiritual values which may not necessarily be related to an organized church or religious institution.[51]

In terms of spiritual emergency, unfortunately, this does not go far enough and very few have seen it translated into clinical practice, certainly in the UK. This is because a key issue has not been addressed, that of how mental health professionals are trained and what is included in curricula.

Apart from the category 'Religious or Spiritual Problem' there was also effectively a recognition in the DSM-IV that certain spiritual practices can trigger what is known in psychiatry as 'brief reactive psychosis'. Appendix 1, Culture Bound Syndromes, referred to 'Qigong psychotic reaction', drawing on the fact that Qigong (or Chi Gung) induced psychosis is recognized in Chinese psychiatry.[52] Qigong, which can be translated as 'energy cultivation', is a meditative movement practice, similar to Tai Chi. Whilst the DSM-IV does not acknowledge that other spiritual practices, when practiced intensely, can have the same effect, it is tacitly implied. A new version of the DSM, the DSM V, is due out May 2013.

Introducing the new category 'Religious or Spiritual Problem' into the DSM in some respects marked a step forward. But there are also dangers. There can be a tendency to want to try to differentiate between what are seen as pathological states, so-called 'regressive', and those states that offer the possibility for growth and healing. All the transpersonal thinkers we looked at in the last chapter take the view that non-ordinary states, including psychosis, offer the potential for healing and growth, if supported in the right way.

Even if we accept that some states are pathological, the reality is still that in spiritual emergency the two are so frequently inter-mingled, with aspects of both the pathological and non-pathological being present, that to try to separate them out and say this person is psychotic, whereas this person is in spiritual emergency, is not so helpful.

Assagioli was perhaps the first to identify this issue. Having made a distinction between the two groups, he then acknowledges that: "In some cases the treatment is complicated by the fact that there is an admixture of 'regressive' and 'progressive' symptoms."[53]

In other words individual experience, my experience, your experience, does not fit neatly into boxes, the boxes of pathological and non-pathological. By the time Assagioli gets to the end of the same paragraph he has had to admit that: "… a careful analysis shows that **most** of those who are engaged in the process of self-actualization are to be found with remnants of this kind [regressive] …"[54] (my bold)

In one case, somebody I met, who had been given a psychiatric diagnosis in the past, went back and had his medical records changed to acknowledge the aspect of spiritual emergency. It can be very healing to get a second opinion, for the spiritual emergency features to be recognized. This is not to deny any psychotic element, but to see the diagnosis as a case of both/and, as opposed to either/or.

Isabel Clarke is an NHS Consultant Clinical Psychologist and one of the Directors of the Spiritual Crisis Network. She has consistently argued that asking whether someone is psychotic or in spiritual crisis is asking the wrong question. Her clinical practice and research has been based more on seeing such transliminal states of consciousness as being part of a whole spectrum. Experiences anywhere along the continuum have the potential to be spiritually transformative.

The Value of Research

Increasingly research is available to corroborate anecdotal evidence about spiritual emergency. To give one example, Caroline Brett's chapter in Clarke's *Psychosis and Spirituality: Consolidating the New Paradigm* (Wiley-Blackwell 2010) reports on her Ph.D. findings. Brett touches on the fact that the experience of spiritual emergency may not be that different in itself from the experience of so-called psychosis. What differs is the context, the outcomes and, of course, the labels.

I asked a related question at a conference organized by Norfolk & Waveney Mental Health Care Partnership NHS Trust (UK) when I spoke on *Psychospiritual Crisis:*

Where Mysticism and Mental Health Meet. What makes the difference? What makes the difference between one person being shattered by their experience and ending up in hospital, possibly repeatedly? What makes the difference between that and another person coming through their experience stronger, with a clearer sense of purpose and meaning, more integrated psychologically? What is it that enables one person to flourish and causes another to flounder? The conclusion I came to, based on my own experience of having both floundered and flourished following two very contrasting periods of spiritual emergency, was that two factors in particular are crucial. One is the individual's frame of reference, the lens through which they view their own experience, and the other is whether their experience is validated or pathologized. (The full talk is available as an MP3 — see Resources). This is indeed what the research of Clarke, Jackson and others reported in *Psychosis and Spirituality* (Wiley-Blackwell 2010) is beginning to show.

The Flirting of West with East

The acknowledgment in the DSM of 'Qigong psychotic reaction' appears to have been as a result of difficulties encountered amongst the Chinese immigrant population in the USA. Overall, however, psychiatry seems woefully slow to bring together Eastern and Western understanding. It lags far behind what the population at large has taken on board.

Yet any scientific or medical attempt to research and map the territory of spiritual emergency has to bring together Eastern and Western perspectives. Ancient religious traditions such as Hinduism and Buddhism have such a rich and detailed understanding of non-ordinary states of consciousness, including enlightenment, and highly complex systems for helping adepts progress towards these, that they cannot be ignored. At the same time, Western psychology has a great deal to offer, from William James's first charting the *Varieties of Religious Experience* through to Grof's work and beyond. Those who have contributed the most, such as the Carl Jungs and the Joseph Campbells, drew heavily on both Eastern and Western schools of thought and successfully achieved a marriage of the two.

Without an appreciation of ourselves as beings of energy, as much as physical beings, we would not be able to begin to make sense of spiritual emergency. One of the most fundamental needs during spiritual crisis is to ground the vast influx of energy. Any professional, whether mental health clinician or other, will not be able to help with this all important grounding if they have not grasped the basic principles, in other words, if they do not incorporate Eastern understanding into their Western approach. Therapies such as Shiatsu, a cross between massage and acupuncture, which come to us from Japan and China, are so helpful for spiritual crisis, because they are based on this understanding of our bodies as energy bodies. They focus on balancing those energies. Acupuncture works on the same principle as Shiatsu, but the particular gentleness of Shiatsu is especially suited to spiritual emergency.

In today's environment, with multi-ethnic and multi-faith clinical teams as well as client groups, this broad knowledge base is all the more important. John Nelson's *Healing the Split*[55] is a good example of this synthesis of Eastern and Western approaches, drawing as he does on the seven chakras as a framework for exploring the relationship between spirituality and mental health.

The Good News

Perhaps nowhere is the integration of ancient Eastern spiritual understanding into Western medicine as remarkable nor as widespread as in the field of Mindfulness. Jon Kabat-Zinn's Mindfulness Based Stress Reduction program[56] and research has spearheaded the introduction of Mindfulness into the mainstream, offering relief for a whole range of physical and mental health conditions. On the mental health side, most of the research in the UK has focused on relapse prevention for depression,[57] although there is also some research into its use with so-called psychosis.[58]

I have been teaching the classic 8-week Mindfulness course since 2007 and have experienced first-hand how beneficial people find it. For me, this only serves to confirm the positive research outcomes. Such results, along with discovering how helpful I found Mindfulness in coping with the worst aspects of my spiritual emergencies, gives me great confidence in advocating Mindfulness for spiritual emergency.

Who defines reality?

Amazingly, when European explorers landed on the southern tip of South America their vast ships were so far beyond what the natives could conceive of, they were actually invisible to them. They were unable to see them, although they were able to see smaller landing crafts. Rather than see what they were looking at, they saw what they were looking for.[59]

In the same way, before Roger Bannister broke the four-minute mile, it was not thought physically possible for a human being to run a mile in less than four minutes. Within seven months of his record 37 other athletes had succeeded in matching it.[60] They now believed it was possible. What was previously thought of as impossible was now known to be achievable; the consensual reality of the athletics world had expanded.

Consensual reality means that the consensus on a particular subject becomes reality, it defines reality. In the realm of spiritual emergency, if we are optimistic, we can nsensual reality starting to expand. We have seen how Kundalini, near-death and out-of-body experiences and others are now becoming more readily recognized and accepted. The consensual reality of the medical world is beginning to broaden to include these phenomena. With near-death and out-of-body experiences, this is partly thanks to extensive research that has been carried out and also because modern resuscitation methods mean that far more people are having these experiences in hospital than ever before.

Let us hope that in time medical consensual reality will expand sufficiently to fully include all the different types of phenomena encountered in spiritual emergency, including past-life material and spirit attachment. At the grass roots level, what helps are studies like that conducted by Kaia Nightingale in Canada and the many interviews that Hazel Courteney has recorded with international researchers and scientists, reported in her various books.

At the academic level, research and publications furthering this shift in the UK include the latest edited volume brought out by Isabel Clarke, with its many contributors. Also helping is the work of the Special Interest Group on Spirituality and Psychiatry of the Royal College of Psychiatrists, with texts such as *Spirituality and Psychiatry*. In America the work of clinicians such as David Lukoff, Co-President of the American Association for Transpersonal Psychology, has been influential, along with a whole host of transpersonal bodies, such as the Institute of Transpersonal Psychology and the International Transpersonal Association, to name just two.

A Word of Caution

Whilst the research and evidence base supporting an understanding of spiritual emergency and its various manifestations is out there and available, I would not want to give false hope to anyone going through or recovering from such a crisis. It is early days. More research is needed in order for mainstream mental health services to really take the phenomenon of spiritual crisis on board. I have, before now, encouraged people to take research papers into meetings with psychiatrists, for example, downloaded from the website of Royal College of Psychiatry's Special Interest Group on Spirituality.

In Part I of the book, What is Spiritual Emergency?, we have touched on the parallels between individual crisis and global crisis and we have heard what spiritual emergency can be like from those who have been through it. We have also explored the transpersonal lineage, the pioneers of psychology and their legacy that has today made possible the kind of research being conducted. In Part II, Spiritual Emergency Through the Ages, we will look at prominent individuals and their experiences of spiritual emergence and emergency.

Spiritual Emergency
Through the Ages

CHAPTER 5

The Mystics

In the next few chapters, we will explore spiritual emergency down through the ages, from the early mystics to well-known people today. Spiritual emergency has been around since time immemorial. Figures such as the Buddha and Jesus faced trials and periods of intense crisis. In the process of gaining enlightenment, the Buddha was beset by the dark forces of Mara, as he sat under the Bodhi tree, meditating. Jesus spent forty days in the wilderness of the desert, facing his own demons, before he was able to begin his ministry. These periods of deep crisis presaged major spiritual breakthroughs and it was no different for the mystics whose experiences we will explore.

In reading about their lives, it is sometimes difficult to relate to them today. The severe, self-imposed physical suffering they subjected their bodies to, can seem alien to us. But if we are able to look beyond this, we find extraordinary examples of spiritual emergency and awakening, examples that are nothing short of inspirational.

YESHE TSOGYAL, LADY OF THE LOTUS-BORN

Yeshe Tsogyal was a remarkable woman. Born in Tibet in the Eighth Century, she attained full enlightenment, is credited with many miracles and played a key role in establishing Buddhism in Tibet. Much of what we know of her comes from *Lady of the Lotus-Born: The Life and Enlightenment of Yeshe Tsogyal*, a text written by two of her contemporaries that is part biography and part autobiography. Her life story is also the fascinating history of how Buddhism was introduced to Tibet from India and how it came to replace the early Tibetan shamanic spirituality called Bon. At that time the empire of the Tibetan kings reached far into present-day China, into Central Asia and extended over the whole Himalayan region.

King Trisong Detsen, who reigned c.755-797, invited Padmasambhava, the great Indian Buddhist master and mystic, to help spread the teachings in Tibet. Also known as the Lotus-Born, Padmasambhava is now revered as a Buddha. At the age of sixteen, Yeshe Tsogyal became his disciple and consort. Together they practiced the highly complex tantric path, according to which it is possible for true adepts to achieve Buddhahood in a single lifetime. In the West today, tantra tends to be associated with sacred sexuality, using sexual energy as a path to awakening. In actual fact that is only one aspect of the tantric path as practiced in Tibetan Buddhism.

Before she became Padmasambhava's consort, Yeshe Tsogyal had already seen quite a bit of life. Born with auspicious signs and omens, her parents were aware that this was no ordinary child. They turned away all who requested her as their bride until two lords vied for her. To settle the matter, Yeshe's father chased her out of the house, saying that whoever managed to catch her could marry her. When caught by the Lord of Kharchu, Yeshe did her best to resist and was savagely beaten. She managed to run away and lived wild for a time, until the Lord of Surkhar heard of her whereabouts and sent a huge army to find her. This time Yeshe was put in chains. War between the two lords seemed imminent. The king came to hear of these events and of Yeshe's outstanding qualities and decided that she would make a fitting queen for himself. The conflict was averted thanks to Yeshe having two sisters, who were each married to one of the lords.

The king now had a total of five wives. Not long after he had celebrated his marriage to Yeshe by feasting for three months, Padmasambhava arrived. The king requested the powerful secret teachings and Padmasambhava told him that in order to spread them throughout Tibet an exceptional woman would be needed. The king, in awe as Padmasambhava miraculously appeared in the form of the tantric Buddha Vajradhara, offered him Yeshe. We are told master and consort then went to a place called Chimphu Geu where they practiced in secret.

Yeshe's life became one of total dedication and devotion to the teachings, the practice and her guru. Years of intense meditation and tantra, along with the transmission of teachings and energy from Padmasambhava resulted in her achieving stage after stage of enlightenment. Through the arduous years of practice, she displayed incredible courage, determination, endurance and faith.

One of the many miracles she is credited with is bringing back to life a soldier who had died of sword wounds. Many students gathered around Yeshe, as she travelled the length and breadth of the Tibetan kingdom teaching, building spiritual communities, giving prophecies and concealing treasures of the doctrine to be discovered by future generations.

In terms of our theme of spiritual emergency, there were several crucial times of crisis during Yeshe's intense spiritual life, each of which marked an important turning point. As she was able to move through each critical crisis, so she advanced towards full enlightenment, step by step.

Padmasambhava had given Yeshe eight instructions or 'precepts', which she vowed to follow in her practice. It was as she focused on each of these in turn that she met with crisis after crisis. The first of these involved achieving mastery over body temperature, known as the inner heat of *tummo*. Up in the snowy mountains of Tibet, Yeshe meditated for a year, wearing only a piece of cotton cloth. But the warmth of the *tummo* did not come and she found the gales, frost and cold almost unbearable. Because of her vow she persevered. We are told her whole body was covered in blisters. Her breathing painful, she was on the point of death, when at last she prayed to Padmasambhava. As the energy and warmth of the *tummo* began to rise, she felt her prayers answered.

> My body, which beforehand had been completely frozen, was transfigured through and through, like a snake sloughing off its skin.
>
> *LADY OF THE LOTUS-BORN* [61]

The image of the snake shedding its skin, with its symbolism of rebirth, is often associated with spiritual emergency. Yeshe then began to focus on the next challenge, which was to overcome the need for physical nourishment.

> At that time, I had not even a single barley seed left, and so I performed my meditation taking stones and water for my food and drink. After some time…[my] legs could no longer support the weight of my body. I could not lift my head, and I had difficulty in breathing… My mind, too, grew very feeble. Gradually my condition worsened, so that in the end I was on the brink of death.
>
> *LADY OF THE LOTUS-BORN* [62]

Again she prayed to Padmasambhava and this time she had a vision of a Goddess, following which she felt her body become "as strong as a lion". She very simply tells us that in her meditation she "realized the ineffable truth" and was then ready to move on:

> I thought then that the time had arrived for me to go unclothed and to take the air for my food. And so, for one whole year, relying solely on the air for sustenance, I practiced naked.
>
> *LADY OF THE LOTUS-BORN* [63]

This phenomenon of not needing food is known in the 21st Century with stories of such practitioners appearing from time to time in the Press. It is, as Yeshe discovered, a highly dangerous practice. At first she was doing fine, but then, as a result of doubting she again found herself close to death. This time when she prayed to Padmasambh?

he appeared before her, telling her she was taking things to excess and should use the essence of medicinal plants and herbs to restore her body.

And so the stages continued, each one as demanding in its own way as the last. In one she made enormous personal sacrifices in order to help others in need. In another she was beset with horrendous visions and tortures by the spirit world, including sexual taunting and molesting. Annoyed at their failure to disrupt her concentrated meditation, the same spirits sent storms, floods and plague to the area, inciting the local people to attack her. In all of this, the Lady of the Lotus-Born never wavered. She is said to have achieved the state in which the body is neither affected by death nor aging.

She was, however, still not fully enlightened at the time of her final crisis, triggered when Padmasambhava came to take his leave from this world. The sacred bond between guru and disciple was all important to Yeshe. The two had been inseparable, practicing together for years. Yeshe had received many teachings, much encouragement and guidance, and could never have made such outstanding progress without Padmasambhava's help. Indeed, he had saved her life several times. She was distraught at the thought of losing him and tells us:

> I struck my body on the ground, tearing my hair, clawing at my face, rolling on the earth. And I begged him:
>
> Woe and sorrow! Lord of Orgyen!
> Will you leave Tibet an empty land?
> Are you taking back your light of love?
> Do you cast aside the Buddha's teaching?
> Will you throw away the people of Tibet so heedlessly?
> Are you leaving Tsogyal with no refuge?
> O look on me with pity!
> Now, now, look at me!
>
> *LADY OF THE LOTUS-BORN* [64]

Her imploring continued and still she battered her body against the rocks, until she drew blood. Finally, having poured out her grief in song, and having received miraculous signs of Padmasambhava's continued and everlasting presence, she says:

> I gained a fearless confidence: the nest of hope and fears fell to nothing, and the torment of defiled emotions was cleared away. I experienced directly that the Teacher was inseparable from myself.
>
> *LADY OF THE LOTUS-BORN* [65]

This was key, the realization that they could never be parted in reality, that time and space were no obstacle, that teacher and student were one. Over the following three months she had ample proof of this, receiving many instructions, predictions and counsels from him, despite his physical absence. So, with signs and reassurance from Padmasambhava himself, Yeshe was able to move through this crisis of separation from the teacher and went on to gain full Buddhahood.

Still today, in the process of separating from a teacher, some go through protracted periods of spiritual emergency. This is reflected in the earlier mentioned DSM, in which the category 'Religious or Spiritual Problem' includes reference to difficulties in separating from a spiritual teacher. Yeshe's despair, followed by the insight that came with her breakthrough, is just as relevant today as it was in Eighth Century Tibet.

Separation from the master is only one aspect of the guru-disciple relationship that can lead to spiritual emergency. Any rupture in that rapport, especially when due to the unethical behaviour or sexual misconduct of the teacher, can give rise to spiritual emergency. When the teacher-student relationship is healthy, however, and we have that special bond, we can know that, in times of real hardship, our teacher will be available for us. We can call on them across time and space and they will be there to support us. This can indeed give rise to the fearless confidence Yeshe spoke of. Her life story is a testament to the potential that can be unleashed and the spiritual mastery that can be achieved, when the guru-disciple relationship is sound.

ST. TERESA OF ÁVILA

St. Teresa was a Spanish mystic, born 1515 in Ávila, some 70 miles north-west of Madrid. Entering the Carmelite Order as a young woman, she suffered from severe and mysterious fainting fits for much of her adult life. Eventually these changed into experiences of rapture and in the latter part of her life she founded many convents and wrote an influential body of mystical literature.

What is relevant for us is that she suffered the longest, severest spiritual emergency on record. From the age of eighteen, for twenty-five years she experienced strange attacks that could not be diagnosed. These left her semi-conscious, her body wracked with pain and paralyzed, until, following a breakthrough spiritual experience, these seizures transformed into ecstatic raptures.

We can trace the roots of Teresa's spiritual emergency and awakening back even earlier, to her childhood. When she was 12 years old her mother died. As we saw earlier, bereavement or loss of any kind can trigger spiritual emergency. Writing about it in her autobiography, Teresa said:

… when I began to realize my loss, I went in my distress to an image of
Our Lady and, weeping bitterly, begged her to be my mother.

THE LIFE OF SAINT TERESA OF ÁVILA BY HERSELF [66]

This was a powerful spiritual act at the tender age of 12. By the time she was 16, however, she was a typical teenager. Her widowed father, concerned that she was showing too much interest in romantic novels and boys, sent her to be educated at a local convent. After a spell back at home due to poor health, she entered the Carmelite Order, becoming a nun, initially against her father's wishes. Her health, however, was already not good and soon went from bad to worse. Teresa says:

My fainting fits began to become more frequent, and I suffered from
such pains in the heart that everyone who saw them was alarmed… I
had also many other ailments. I spent my first year, therefore, in a bad
state of health.

THE LIFE OF SAINT TERESA OF ÁVILA BY HERSELF [67]

Her mention of severe pain in the heart area is fascinating, given what Jack Kornfield, the Buddhist teacher and writer, has to say about this. He tells us that sometimes, on intense meditation retreats, participants are convinced they are having a heart attack, because of the severe pain in the chest area. To their consternation, he sends them back to their meditation. [68] The Eastern understanding of this, as we have seen, is that we have seven energy centers positioned up through the body, along the spine, called chakras. As blockages are cleared in the heart chakra it can be physically painful. Spiritual practice, whether meditation, or prayer and devotion in Teresa's case, can create this freeing up of the heart chakra.

Later Teresa came to recognize these strange states as spiritual seizures or raptures, but at the time neither she nor those around her understood what her strange afflictions were. She wrote:

…my condition was so serious that I was usually semi-conscious and
sometimes lost consciousness altogether.

THE LIFE OF SAINT TERESA OF ÁVILA BY HERSELF [69]

On a personal note, having experienced very mild equivalent 'episodes', had I not previously read Teresa of Ávila's work, I would have either been very frightened or thought I was seriously ill. Eventually, due to these strange, undiagnosed fainting fits, Teresa was obliged to leave the convent for a year. Her father sent her for a three-month 'cure', but the primitive medieval medicine left her in a very sorry state:

> [I] suffered the greatest tortures from the remedies they applied to me which were so drastic that I do not know how I endured them. In fact though I did endure them, they were too much for my constitution.
>
> *THE LIFE OF SAINT TERESA OF ÁVILA BY HERSELF* [70]

Shortly after she suffered such a severe attack that everyone presumed she would die.

> I had an attack which left me insensible for almost four days. They gave me the Sacrament of the Extreme Unction and in every minute of every hour thought that I was dying… For a day and a half a grave was left open in my convent, waiting for my body.
>
> *THE LIFE OF SAINT TERESA OF ÁVILA BY HERSELF* [71]

Teresa seems almost to have come back from the dead. Here we see the powerful theme of death and rebirth so common in spiritual emergency. Those going through such crises often feel or fear that they are dying. Sometimes this is because the physical effects are so strong and sometimes, as we have seen, it is because the ego feels so threatened as it is extinguished, however temporarily, that we believe we are physically dying.

The very severe physical trials that Teresa went through help us gain a better understanding of present day accounts of spiritual emergency. It helps to put in context the experience of somebody like Jennifer, whose extreme physical pain and illness you read about earlier. We can see this is in terms of the purification process of mind, body and soul that accompanies spiritual awakening. In spiritual emergency it can, however, be very harsh on the body.

The spiritual emergency came to a head for Teresa when she was thirty-nine. She had a breakthrough experience when walking down the corridor of the convent one day. On seeing a particular statue of Jesus, she suddenly experienced the full extent of his suffering when he was flogged, fell to the floor weeping and in that moment felt the Christ Consciousness enter her. From then on her seizures turned to ecstatic raptures, although the actual physical states hardly varied at all from before. Her body temperature would drop, her breathing would slow right down and she would be unable to move a single part of her body, despite being able to see and hear what was going on around her. Again, on a personal note, what I found was that on occasion I was able to move my arms, legs or torso, but with incredible effort. Today we can speak of such experiences using different language and with the added benefit of Eastern teachings. My understanding of what was happening to me was that a powerful trans-mission of energy was taking place. But can we separate out incoming energy from the upsurge of our own inner energies?

Teresa also describes how, at other times, her body would feel incredibly light and on a very few occasions she was actually seen to lift off the ground. The phenomenon of levitation is not unknown and in Chapter 10 you can read a modern day account of Kimberley's experience of it, when she physically lifted off her bed, whilst going through spiritual emergency. Teresa's chief concern, however, was the inconvenience of the raptures coming over her in public.

> … [they were] likely to arouse considerable talk… once in particular during a sermon — it was our patron's feast and some great ladies were present — I lay on the ground and the sisters came to hold me down, but all the same the rapture was observed.
>
> *THE LIFE OF SAINT TERESA OF ÁVILA BY HERSELF* [72]

One of the features that can be present during spiritual emergency is sometimes mis-taken for grandiosity. Indeed psychiatrists refer to this as delusions of grandeur, as we saw earlier. Teresa gave us an explanation for this:

> I know by experience that the soul in rapture is mistress of everything … It perfectly well sees that this is not its own achievement, and does not know how it has come to possess such a blessing. But it clearly realizes the very great benefit that each of these raptures brings. No one will believe this who has not experienced it; and so people do not believe the poor soul … when they suddenly see it aspire to such heroic heights. For now it is not content to serve the Lord in small things, but wishes to do so in the greatest way it can.
>
> *THE LIFE OF SAINT TERESA OF ÁVILA BY HERSELF* [73]

Emma Bragdon considers that probably what Teresa experienced for years was a series of dramatic Kundalini awakenings. As we have seen, in Eastern understanding, Kundalini energy resides dormant at the base of the spine. With the process of spiritual opening this energy moves up through the chakras. It very often encounters blockages or resistance, creating a burning sensation of heat as it clears these. Once Teresa's spiritual awakening was complete, once her body, mind and soul had been purified, the attacks stopped.

The fact that her fainting fits were accompanied by fevers is, in fact, suggestive of the intense heat that can be generated by the Kundalini energy. Teresa also talks of a rushing sensation in her head and especially the crown of her head. Mystified by this, she comments that she has heard this is where the soul resides. Today, with the knowledge of Eastern mysticism that has found its way to the West, we can assume that the sensations were in the crown chakra. This too is in line with Kundalini awakening.

In her later years, once the seizures had turned into raptures, Teresa accomplished remarkable things, including establishing eleven convents and writing an influential body of work. She was a classic case of 'in the world, but not of the world'. In Part III we will look at the different stages of processing spiritual emergency. We will see how fulfilling our spiritual calling depends on integrating the experience and being able to go back out into the world effectively. Teresa is a very good example of someone who managed to accomplish that.

ST. JOHN OF THE CROSS

John of the Cross and Teresa of Ávila can be thought of as twin saints. Born only thirty miles and twenty-seven years apart, John, the younger of the two, considered Teresa his spiritual mother. It may have been her writings that inspired and encouraged him to write also. They met not long after he was ordained, when she was fifty-two. Impressed by him, Teresa asked John to help her with the reform of the Carmelite Order. She was in the process of taking it back to its roots of simplicity and contemplation. He willingly became confessor in one of her nunneries.

In 1577, however, the traditional Carmelites, having more or less outlawed Teresa's reformed branch of the Order, imprisoned John in a windowless cell in Toledo, on a diet of bread and water, with frequent flogging. He managed to escape nine months later, but not before having written, amongst others, the poem 'The Dark Night of the Soul'. In the darkness of his cell, he had committed the lines to memory, until a new and kinder guard gave him writing materials and occasionally allowed him out into the sunshine. The long periods of darkness in his cell, with a rare glimpse of light, followed by liberation, mirror the inner spiritual journey he wrote about. As is so often the case in spiritual emergency, external events and the unfolding of our inner process seem to synchronistically mirror each other. The book *The Dark Night of the Soul*, for which John is so well known, came later. It is a line by line commentary of the poem and is aimed at guiding and helping people on the spiritual path.

When the book first became available in English, the Spanish title 'Noche Obscura del Alma' was translated as 'The Obscure Night of the Soul'. This helps us to grasp more easily the essential quality of the 'obscure night', its mystery, its fathomless depths, its defying rational understanding. Feeling confused and lost is central to an experience of the 'obscure night'. There can be a catastrophic sense of loss, either because of the emptying of the 'self' or because of the feeling of separation from the Divine. Feeling abandoned by God is quite common. Here is how St. John of the Cross puts it:

> The self is in the dark because it is blinded by a light greater than it can bear. The more clear the light, the more does it blind the eyes of the owl, and the stronger the sun's rays, the more it blinds the visual organs, overcoming them by reason of their weakness, depriving them of the power of seeing... As eyes weakened and clouded suffer pain when the clear light beats upon them, so the soul, by reason of its impurity, suffers exceedingly when the Divine Light really shines upon it. [74]

In his writing, John distinguishes between two stages of the 'dark night'; the dark night of the senses and the dark night of the spirit. The dark night of the senses is marked by the letting go of attachment to sensory pleasure. In the dark night of the spirit, the intellect, memory and will are obscured, since the Divine cannot be understood through these. The whole process is one of emptying of self. Then pure love can flow through the soul and it can experience union with the Divine, the metaphorical coming of dawn after the night.

What happens is that we so often fall into self-doubt and self-blame that the experience can become even more painful. We think that we have gone horribly wrong somewhere, because we imagine it is all down to us. We think we should be able to 'do' something about it. We fail to see the secret, mysterious workings of the 'dark night'. If we can be with, sit with, live with the uncertainty, the confusion, the not knowing, the journey will be easier. If we can resist trying to figure it all out, trying to control it, willfully trying to make everything okay, then the experience will be that much less painful. We stand a chance of moving through, rather than creating resistance, of moving through to something new and infinitely better.

This is the new dawn that John speaks of, that emerges after the 'obscure night'. The poem shows this clearly. The book, however, stops short, the commentary covering less than half the poem. Many of John's works were left unfinished and it is not clear why he stopped writing, given that he lived many years more.

I have included the poem here because it helps to illustrate how John's 'dark night' has to some extent been misrepresented, and also because it shows so clearly the parallels with other mystical traditions. The exquisite, devotional lines of some Sufis, Islamic mystics, such as Rumi and Hafiz, come particularly to mind. They also personify God as the Beloved.

The Dark Night of the Soul

On a dark night,
Kindled in love with yearnings – oh, happy chance!
I went forth without being observed,
My house being now at rest.

In darkness and secure,
By the secret ladder, disguised – oh, happy chance!
In darkness and concealment,
My house being now at rest.

In the happy night,
In secret, when none saw me,
Nor I beheld aught,
Without light or guide, save that which burned in my heart.

This light guided me
More surely than the light of noonday
To the place where he (well I knew who!) was awaiting me –
A place where none appeared.

Oh, night that guided me,
Oh, night more lovely than the dawn,
Oh, night that joined Beloved with lover,
Lover transformed in the Beloved!

Upon my flowery breast,
Kept wholly for himself alone,
There he stayed sleeping, and I caressed him,
And the fanning of the cedars made a breeze.

The breeze blew from the turret
As I parted his locks;
With his gentle hand he wounded my neck
And caused all my senses to be suspended.

I remained, lost in oblivion;
My face I reclined on the Beloved.
All ceased and I abandoned myself,
Leaving my cares forgotten among the lilies.[75]

John's book only covers the first three stanzas of the poem, and deals very much with the trials and tribulations of the spiritual path. The later stanzas speak far more of the kind of rapturous states in which the self is totally surrendered, ecstatic states that Teresa wrote so freely about. It is as if our knowledge and understanding of the 'dark night' has been limited to the first three stanzas, the last five being neglected and forgotten.

In terms of spiritual emergency today, very few people recognize they are going through a 'dark night'. As the American psychiatrist Gerald May says, in his book 'The Dark Night of the Soul', we are very likely to put our mental state down to all sorts of other things. Conversely, if we think we are going through one, we probably are not. A large part of why we might miss it is because so many today do not consciously think in spiritual terms, including our mental health professionals; it is much easier to assume or conclude that it must be depression. Psychiatrists such as May, who have such a sensitive awareness of the 'dark night', are rare indeed.

The lives of St. Teresa of Ávila and St. John of the Cross seem particularly relevant to our understanding of spiritual emergency today. Experiences like Teresa's that bring bliss, unbounded love and insight, can take us into the realm of 'mystical psychosis', whilst the sense of loss and desolation can make the 'dark night' look and feel remarkably like depression. If spiritual crisis can lift us up into the heights of mysticism, and mystical psychosis, and take us down into the depths of the dark night of the soul and depression, then St. Teresa and St. John are perfect examples of two apparently contrasting ways in which spiritual emergency can manifest. Both, however, are processes of purification, clearing us out for a higher, and deeper, spiritual engagement.

RAMANA MAHARSHI

Ramana Maharshi, affectionately known as Sri Bhagavan, meaning 'the blessed one' or 'the divine', was an Indian sage. He became known and loved in the West, as well as India. Venkataraman, his original name, was born into a Hindu family in a village of Tamil Nadu, South India, in 1879. This was the era of Queen Victoria and the colonial British Raj and Venkataraman's father was keen for his sons to learn English, so that they could enter government service.

At the age of 16, Venkataraman had an experience of spontaneous awakening. Never particularly studious, his school studies then lost all meaning for him and soon afterwards he felt irresistibly drawn to the sacred mountain of Arunachala at Tiruvannamalai, in Tamil Nadu. He left home and spent the rest of his life living on or around the mountain and the temple complex there. For several years he maintained complete silence, spending days, even weeks, absorbed in deep meditative states and depending on others for food. Gradually word spread of this Self-realized master and people began to gather around him.

Sitting in his presence, devotees would soak up the inner silence and stillness he radiated. Sri Bhagavan considered silence to be the purest form of his teaching. Later, when he did at times speak, his teachings are described as flowing from his direct experience of consciousness as the only existing reality. He encouraged self-enquiry and the use of the question 'who am I?' as the path to liberation. His approach was essentially that of Advaita Vedanta, the school of Hinduism that focuses on non-duality.

As the number of devotees grew, so did the ashram around him, eventually including a hospital, library and post office. Notable visitors included Paramahansa Yogananda, author of *Autobiography of a Yogi*, who took Kriya yoga to the West, as well as Krishnamurti, the philosopher, who experienced the awakening of his Kundalini energy in Sri Bhagavan's presence.

There are several aspects of Sri Bhagavan's experience that are relevant to our exploration of spiritual emergency, including his spontaneous awakening and his attitude towards his body afterwards. His awakening is interesting because, like others, he too keenly felt the fear of dying. It was this sudden fear, which came out of the blue one afternoon when he was alone in the house, which initiated his enquiry. "What is it that dies?" he asked. "Who am *I*?" He lay on the floor like a corpse, mimicking death. As the experience of realization unfolded, he discovered that consciousness survives physical death. As Sri Bhagavan stepped towards his ego's fear of death, the ego dissolved, as did the fear of dying.

As we have seen, many people when going through a crisis of spiritual awakening experience this fear of death, very often confusing the ego's death with physical death. What is exceptional in Sri Bhagavan's case is that the ego dissolved permanently, as opposed to temporarily. Not only did he see through the ego as being non-existent, he also experientially understood that we are not our thoughts, nor are we our bodies. We only inhabit them temporarily. His behaviour over the following months was therefore not that surprising. He was so absorbed in this new state of altered consciousness that he showed a total disregard for his body. He was incapable of looking after his most basic physical needs. If someone had not actually put food into his mouth in those early days at Arunachala, he would have starved to death. Luckily some of those attending the temple understood what was happening and that he needed protection and looking after and took it upon themselves to do so.

At that time, as he never washed, his body was covered in dirt, his hair matted and his nails so long and curved that he could not use his hands. Urchins took to throwing stones at this youth who was not much older than themselves, finding it amusing to try to disturb him. One went so far as to urinate on him, but he never stirred when deep in meditation. Sri Bhagavan moved to an underground cellar to avoid this, but there the vermin fed on him, such was the depth of his absorption. When he was carried out from there, his legs were covered in sores, bleeding and oozing pus.

After maintaining silence for three years, when Sri Bhagavan first tried to speak the words did not come out clearly. It took some time for his speech to return to normal. He was also unable to walk and his lack of food intake seriously disturbed his bodily functions.

This is a very extreme example of someone going through the awakening process, who became unable to look after himself. One of the most important things when dealing with spiritual emergency is to keep the person safe, both from themselves and from others. Sri Bhagavan's self-appointed attendants were, at the same time as looking after him, able to respect and support his incredible awakening. They were able to trust the process.

In due course, Sri Ramana came to take care of himself again. This is how Somerset Maugham, the writer, described Sri Ramana in his book *Points of View*, when he visited the ashram in 1936:

> Though he wore nothing but a…loin-cloth (what his biographer somewhat inelegantly calls a cod-piece) he looked neat, very clean and almost dapper. [76]

After his initial awakening, a couple of months had passed before Sri Bhagavan set out for Arunachala. During that time he had taken to visiting a local temple often. He said:

> …in the presence of the images of the Deities…[I] would be beside myself with emotion and would be tear-filled. I did not know what that agony or emotion was due to… My whole body was flooded with this emotion and had a burning sensation. [77]

Again we find intense emotions and this feeling of extreme heat, which for Sri Bhagavan only cooled down when he sat in meditation. When he arrived by train at Arunachala, some weeks later, he hurried straight to the temple. It would normally have been closed at that early hour of the day, but the doors were open and he was able to go right into the inner sanctuary. With the words "Father, I have come according to your bidding, I offer myself to you", he felt total peace. Both the emotional upheaval and the burning sensation were gone.

On another occasion, years later, he experienced something, which is reminiscent of Teresa's experiences. One morning, walking back with a few others from having taken an oil bath, Sri Bhagavan began to feel faint.

> Suddenly the view of natural scenery in front of me disappeared and a bright white curtain was drawn across the line of my vision and shut out the view of nature… On experiencing this I stopped walking lest I should fall…the bright white curtain had completely shut out my vision, my head was swimming, and my blood circulation and breathing stopped. The skin turned a livid blue… Vasudeva Sastri took me in fact to be dead, held me in his embrace and began to weep aloud and lament my death. [78]

This state lasted some ten minutes, until Sri Bhagavan felt a sudden shock pass through his body, then his circulation and breathing revived. His skin returned to its normal color. After this, Sri Bhagavan was reportedly more able to be involved in practical matters, such as planning building projects. He maintained his inner state of realization and stillness throughout.

We can learn much from Sri Bhagavan's extraordinary story. Perhaps it could only have happened in India, where the social, cultural and spiritual norms could allow for someone to be physically reduced to such a state and yet be understood to be going through a profoundly important process of spiritual awakening.

Annabel's story, mentioned earlier, is a present day example of the wildly varying responses different norms can give rise to. She went through a powerful experience of spiritual emergence and emergency whilst staying with a family in India. Understanding what was happening, they were able to support Annabel and her experience, enabling her to move through it. This was both empowering and affirming for her. Back in the UK, however, a similar crisis resulted in her being sectioned, that is committed against her will to a psychiatric hospital, which was both traumatic and pathologizing for her. As Carl Jung wrote:

> The life and teachings of Sri Ramana are not only important for the Indian but also for the Westerner. Not only do they form a record of great human interest, but also a warning message to a humanity which threatens to lose itself in the chaos of its unconsciousness. [79]

There is much in the lives of these four mystics of relevance to present day experiences of spiritual emergency. We can begin to get a sense that as long as individuals have been dedicating themselves to the spiritual path, so awakening has been accompanied by spiritual emergency.

CHAPTER 6

The Creatives

In this chapter we will look at those individuals, who although not spiritual masters in their own right, with the exception of Hildegard of Bingen, were deeply in touch with their spiritual life and left an outstanding body of creative work.

Creativity and spirituality are closely related. Julia Cameron, author of *The Artist's Way*, asserts that when we get in touch with the creative energy of God, the Divine or the Source, our own creativity will flow. And it will come from our higher Self or may even feel as if it is coming from outside ourselves. Many poets, composers, writers feel that they are simply just a channel. So let us explore the relationship between spiritual crisis and creativity through the lives of three highly talented individuals, who were nothing short of creative geniuses and who all experienced spiritual emergency.

HILDEGARD OF BINGEN

'A feather on the breath of God' — this is how Hildegard of Bingen, the 12th Century visionary prophet described herself; a beautiful image that captures her sense of surrender to the Holy Spirit. Beautiful images were something of a hallmark of Hildegard, both metaphorically, in her poetic writing, and pictorially, in the stunning illuminations of her manuscripts. Here is a brief excerpt from one of her visions from the *Book of Divine Works*, a vision of Love:

> And I saw as amid the airs of the South in the mystery of God a beautiful and marvellous image of a human figure, her face was of such beauty and brightness that I could more easily have stared at the sun …The figure spoke: …I am the fiery life of divine substance, I blaze above the beauty of the fields, I shine in the waters, I burn in sun, moon and stars. [80]

Better known today for her music and her natural health remedies than for her visions and prophecies, she was an extraordinarily creative woman. Talented and prolific in many different fields, she was a Renaissance Woman long before the Renaissance even happened. In an age when women went unseen and unheard, Hildegard was both highly visible and highly audible. This only makes her all the more astonishing. That she claimed to speak on behalf of God and that her revelations were deemed genuine by the Pope, gave her a voice and an authority that would otherwise have been unthinkable for a woman of her time.

Born in 1098, Hildegard was the tenth child of a wealthy, aristocratic family from Bingen, a town south-west of Frankfurt, Germany. From an early age she must have shown signs of being an unusual child. She tells us, in the semi-autobiographical *Life of Hildegard* by Theoderich of Echternach:

> When in my first formation God roused me in my mother's womb with the breath of life, he fixed this gift of visions in my soul...And in my third year I saw such a light that my whole soul trembled, but because of my young age I could not put it in words. [81]

Possibly because of this and also because she was the tenth child, Hildegard was 'tithed' to the Church at the age of eight. In her words, she was 'offered to God for the spiritual life', which she readily consented to. She was put under the care of Jutta of Sponheim, who became a saint in her own right. When, a few years later, Jutta chose to enter an anchorage, attached to the monastery of St. Disibod near Mainz, in the same area of Germany, Hildegard went with her.

Anchorages were small cells consisting of one or two rooms in which the renunciants were literally walled up in order to dedicate their life to devotion and prayer, in service of God. A window was left open to pass food through. Entering the anchorage would have been accompanied by rituals of death, to the mundane world, and rebirth, to the spiritual life. As events would have it, as more women were attracted to follow and serve Jutta, the anchorage rapidly expanded. By the time Hildegard was 15 there was in effect a second monastery adjoining the original monastery for men. At that point she took her vows as a nun.

Jutta was Hildegard's spiritual guide, mentor and tutor in every way, including teaching her the rudiments of Latin. When Jutta died, Hildegard was chosen by the nuns to be her successor, in effect to become Abbess. Throughout her time with Jutta, Hildegard had visions, confiding only in Jutta. She in turn told Volmar, one of the monks, who was to become Hildegard's scribe and close supporter for many years, until his death. Hildegard referred to her experiences as seeing 'the reflection of the Living Light'.

From the beginning Hildegard associated these visions with 'illness'. They had a debilitating effect on her health, at times leaving her immobilized, physically unable to move, in a way that is reminiscent of St. Teresa of Ávila. There is a suggestion that she suffered from migraines, but this does not fully explain all she experienced. Unlike St. Teresa, Hildegard only ever lost consciousness once. She was keen to stress, in the foreword to *The Book of Divine Works*, that her revelations came when she was fully awake and conscious:

> I saw with the inner eyes of my spirit and heard with my inner ears, in heavenly mysteries, fully awake in body and mind — and not in dreams, nor in ecstasy...[82]

Although accompanied by poor health, Hildegard's spiritual life seems to have deepened and strengthened gradually over many years. Her spiritual emergence did not reach crisis proportions until a few years after Jutta's death.

Now she had her first direct experience of what she called the 'Living Light', which was considerably more intense than the earlier 'reflection of the Living Light'. In her Introduction to *Scivias* she described it like this:

> ...when I was forty-two years and seven months of age, a fiery light, flashing intensely, came from the open vault of heaven and poured through my whole brain. Like a flame that is hot without burning it kindled all my heart and all my breast, just as the sun warms anything on which its rays fall.[83]

Suddenly, she found she could understand and offer new interpretations of sacred texts and she heard a voice, which she took to be the Holy Spirit, telling her to write about what she saw and heard. In accessing the Living Light, she seems to have written, almost dictation style, what she saw and heard. This access to new understanding and knowledge resulted in her three volumes of revelations, the first being *Scivias* (*Know the Ways*), which took her ten years to complete. Here is a snippet from her second work *The Book of Life's Merits*, in which Mercy, personified, replies to Hardness of Heart:

> What are you saying, you creature of stone? The plants give off the fragrance of their flowers. The precious stones reflect their brilliance to others. Every creature yearns for a loving embrace. The whole of nature serves humanity, and in this service offers all her bounty.[84]

Being able to offer new interpretations of holy scripture was, on the other side of the world, in the yogic tradition, one of the tests of whether an adept had attained the highest level of consciousness, of whether they had undergone Kundalini awakening. This no doubt contributed to Gopi Krishna's decision to include Hildegard in a list of those he believed to have experienced Kundalini awakening.[85] Having had his own life-changing Kundalini experience, Gopi Krishna dedicated himself to researching the subject.

I was therefore intrigued to find the following, when reading the Foreword to Hildegard's third volume, *The Book of Divine Works*:

> …when I was sixty-five years of age, I saw a vision of such mystery and power that I trembled all over… [I] set my hands to the task of writing — though I was worn down by so many illnesses, and trembling. [86]

Translators and editors, not experienced in Kundalini awakening or spiritual emergency, tend to assume this is a reference to the fear that Hildegard felt. No doubt she may have experienced some, but the sensation of trembling or vibrating in the body can also, as we know, be strongly associated with the influx of powerful energies and the rising of Kundalini.

A further crisis came when she received Divine guidance to set up a new convent, removing her nuns from St. Disibod. The Abbot Kuno was enraged and flatly refused to give permission. As a result Hildegard became unable to get out of bed. It was only when Kuno tried to move her himself that he became convinced the proposal must in fact be Divine will. It is not clear whether Kuno found her unnaturally heavy, as told in many different miracles, and could not move her, or whether he found her body totally rigid. Either way he declared it a miracle and allowed Hildegard to go ahead.

From her first direct experience of the Living Light, Hildegard's creative outpouring was astonishing. As well as writing the three visionary volumes, these interpretations of the Bible and apocalyptic prophecies were beautifully illustrated. Quite probably she designed the illustrations herself, even if she did not actually execute the painting. She also compiled two medical texts, one on the healing powers of plants, trees, metals, crystals and more, the other on the human body, illnesses and remedies. Both focused on women's health, from menstruation to childbirth, and everything in between. Hildegard further wrote two biographies of saints and a cosmology about the structure of the universe.

She composed nearly 80 songs for worship, despite having had no formal music tuition. These were highly original and very different from the traditional form of Gregorian plainchant, probably because of, or thanks to, her lack of formal training. She saw music and singing not only as a way of praising God, but also as a way of reminding the soul of the sounds of angels singing and thus of the soul's heavenly goal. She wrote of "the sacred sound through which all creation resounds". Perhaps most intriguing of all is what she wrote in the medical text *Causes and Cures* about the music or sound of the cosmos: "In its revolving the firmament emits marvellous sounds, which we nevertheless cannot hear..." [87] Modern science today has confirmed this; NASA has recorded such sound coming off the planets.

Hildegard also composed an opera, *The Play of the Virtues*, which is considered to be the first morality play, the form of theatre that was to become the staple of the Middle Ages. Her numerous letters went to various dignitaries as well as the popes and kings of her day, including the English King Henry II, infamous for orchestrating the murder of Thomas à Becket.

Despite times of spiritual crisis, when Hildegard was unable to function, nor presumably carry out her duties as Abbess or her healing work, she managed nevertheless an impressive creative outpouring. Whether divinely inspired or not, as a woman of the 12th Century her creative genius seems unmatched.

Today, Hildegard's enduring popularity comes mainly from her poetic writing, her herbal remedies and her moving music. Her apocalyptic prophecies have not stood the test of time quite so well. If read carefully, however, there is some fascinating material that strikes a remarkably resonant note with Mayan prophecies of the End of Time. This is from Book Three, Vision 12 of *Scivias*, when Hildegard had a vision of Christ:

> He sat upon a throne of flame, glowing but not burning, which floated on the great tempest which was purifying the world... And again I heard a voice from Heaven, saying to me:
>
> 1. In the last days the world will be dissolved in disasters like a dying man.
>
> These mysteries [the visions] manifest the last days, in which time will be transmuted into the eternity of perpetual light. For the last days will be troubled by many dangers, and the end of the world will be prefigured by many signs ...
>
> 12 . ....the terrors of the elements, the lightnings and thunders and winds and tempests, will cease, and all that is fleeting and transitory will melt away and no longer be, like snow melted by the heat of the sun. And so, by God's dispensation, an exceedingly great calm will arise.[88]

As we approach 2012 and beyond, Hildegard's prophetic words seem to take on a new relevance, some 800 years after they were written.

VINCENT VAN GOGH

If we think Van Gogh, we think sunflowers, we think landscapes of wheat, we think vibrant colors and strong brush strokes. We also think of him cutting his ear off and committing suicide. All this speaks of intensity, from the colors he chose to the sheer volume of work he poured out, from his emotional and psychological distress to how he ended his life. Van Gogh was certainly a man who lived intensely and, as we will see, this intensity reached a crescendo during the last two years of his life.

When we think Van Gogh we do not necessarily think of his deep faith and what he called his 'longing for the infinite'. Van Gogh, the world famous painter, needs no introduction, but I would like to introduce you to Van Gogh, the trainee minister, the missionary, the spiritual seeker. In 1883, he wrote: "…if one feels the need of something grand, something infinite, something that makes one feel aware of God, one need not go far to find it." [89]

Born into a family with a long history of involvement in the worlds of both religion and art, his father was a pastor. Vincent was brought up in a parsonage with the morals of the Calvinist teaching. At 16, thanks to uncles who were wealthy art dealers, Vincent was given a job as clerk in an international art gallery, working first in The Hague, then in the Paris branch, followed by London.

Here, at the age of 21, he seems to have gone through a spiritual crisis and transformation. It was a period that stood out as pivotal in Van Gogh's spiritual journey. For five months, from August 1874, there was a notable gap in his usually prolific letters. He was silent. According to his sister-in-law's memoir, the summer of 1874 was a time of crisis for Vincent and his silence suggests he was going through a powerful inner process. When his correspondence resumed his letters were full of biblical quotations, religious feelings and prayers. Consequently, the period after that summer has been referred to by biographers as one of 'mystic fervour' or 'religious exultation'.

Spiritual emergency often results in a complete change of direction as a person realigns their life and their work with the new found understanding. A true spiritual emergence and emergency is very likely to result in a change of priorities. Whatever crisis Van Gogh went through that summer, it culminated in him deciding to follow in his father's footsteps and train to be a minister. When that route was barred to him because of the rigorous academic demands of the theological school entrance exams, he changed tack slightly and became a missionary. He ministered to the poor miners of the Borinage district in Holland, working tirelessly, giving his own money, clothes and food to those in need and living destitute, as one of them.

Loved and respected by the miners, but dismissed by the Church for his radical behaviour and interpretation of Christ's teachings, Vincent stayed on, sketching the workers and peasants. It was at this point that he decided to become an artist. This was 1879 and Vincent was 26 years old.

Given our interest in Van Gogh's spiritual journey, perhaps the most significant period during the following years, when he painted and studied art, were the months he spent in Drenthe, a moorland region of Holland. The time spent 'on the silent moor', in solitude communing with nature, was healing and regenerating for him. It invoked again the deep connection with the Divine, which he seemed to have felt during his crisis. He could 'feel God high above'. From there he wrote to Theo, his brother:

> What life I think best, oh, without the least shadow of a doubt it is a life consisting of long years of intercourse with nature in the country — and Something on High — inconceivable, "awfully unnameable"…[90]

A few years later Vincent moved to Paris and lived with Theo for two years. Here he met Impressionists such as Gaugin. February 1888 saw Vincent moving to Arles, in Province, where his painting became extraordinary, in terms of how fast he worked and how much he produced.

In and around Arles, Vincent immersed himself in painting nature. He was all too aware of the spiritual connection this brought, given his experience in Drenthe. References to 'excitement' started to appear in his letters, particularly linked with nature: "The excitement, the serious feeling of closeness to nature which guides us — this excitement is at times so terribly powerful…"[91]

He was painting with a feverish fervour, losing himself, 'not even conscious of being at work'. Gaugin came to stay with him and they painted together. Things came to a head on Christmas Eve, when the two men seem to have argued, just after Vincent had heard of his brother's engagement. This could potentially have had serious financial consequences for him, given his total dependence on the weekly allowance Theo was sending him. And, as we have seen, in terms of timing, Christmas and Easter are two very potent times in the Christian calendar. In his distress, Vincent cut off a piece of his ear and presented it to a prostitute in one of the local brothels he and Gaugin frequented, much in the same way the *matadors* of the South would cut off the ear of the vanquished bull and offer it to their beloved.

Vincent did not spend long in hospital and made a good, speedy recovery, but was then hospitalized again at the end of February when the locals, seemingly unprovoked, insisted on his being committed to the nearby asylum. His level of self-awareness was telling. He knew that to protest or get angry would only be used against him as further 'proof' of his 'madness', so he complied. All those years ago he was grappling with the same issues that those going through spiritual crisis today often have to deal with, including at times being forcibly hospitalized.

Much of what Van Gogh described in his letters of this period will resonate with those who have been through spiritual crisis. Writing to Theo just over a fortnight after the ear incident he said:

> Physically I am well, the wound is healing very well and the great loss of blood is being made up, because I eat and digest well. What is to be feared most is insomnia… My suffering from this in the hospital was frightful…'[92]

Comparing two self-portraits, he noticed that his face had become much brighter, whilst his features were unchanged. He described himself as 'a human charged with electricity'. What Assagioli, the psychotherapist, called the 'inrush of spiritual energy' can cause both sleeplessness and this feeling of being charged with electricity, along with being 'unable to attend to the simplest chores', which Vincent also noticed. Writing from the asylum at St. Rémy, he commented on his crises taking a 'religious turn':

> I do not know if this is caused by living in these old cloisters so many months, both in the Arles hospital and here. In fact I really must not live in such an atmosphere, one would be better in the street.[93]

The asylum had previously been a monastery and Van Gogh, possibly psychically very open at the time, seems to have been very sensitive to the energy of the place.

All this can be expected when going through spiritual crisis. Certainly the intensity of Van Gogh's painting during this period suggested someone who was struggling to ground the powerful energy coming through. Later, Vincent felt he had become more patient, the ego more relaxed, the sense of 'self' loosened. He had let go of being attached to outcomes when he was painting. These are some of the qualities one might expect after having moved through a spiritual crisis.

The difficulty for Vincent, as for so many going through spiritual crisis, was that his emotional and psychological wounding was also exposed by the process. And it is this wounding, which if not addressed and healed, can occasionally become highly self-destructive. It is precisely our wounding, which can turn spiritual emergence into the more dangerous spiritual emergency. Ironically it is only a healthy, strong, healed ego that is ready and able to surrender totally to the process of spiritual awakening. With utter clarity, Van Gogh understood the issue here, that it was a question of totally surrendering the ego, and he was not willing to do so:

I believe that recovery will come to him who bravely renounces his egotism and his own will and lets life and death run their own course completely. But that is not proper in my case. I want to paint. [94]

The work that Van Gogh produced during his two years in Provence is some of his most powerful and his most popular. Many art critics have commented on the numinous, transcendent quality of paintings such as *Starry Night*. That Van Gogh was both in touch with the 'infinite' he sought and was able to communicate that on canvas maybe accounts for the enduring popularity of such pieces. How did he get in touch with that in Provence and why did it precipitate such a crisis?

On the one hand he was able to deeply connect with nature, spending whole days, weeks, even months on end, out in nature painting scenes of the different seasons, from sowing to spring blossom and on to harvest. He also had a great deal of solitude, which helps enormously to connect with the transcendent. Van Gogh, himself, acknowledged the monk in him.

Perhaps one of the most powerful factors was the blazing 'midi' sun. Having had personal experience of how the energy of the sun can play a part in spiritual opening, of how it can blast us open, impacting especially on the crown chakra, I can well imagine its effect on Van Gogh. (In my own case, I was on retreat near the Equator and directly experienced the full force of the sun in precipitating a major opening.) We can easily see how working 'under the broiling sun', in his words, could contribute to a very real quickening, a speeding up of the energy pouring through him. We see that reflected in the feverish rate at which he worked.

Van Gogh was highly aware of the sacred energy of the sun, often using it as a symbol or metaphor for God in his paintings. He understood the symbolic realm, so strongly advocated by Jung some forty years later. Often those going through spiritual emergency are able to access this realm and some stay in touch with it afterwards.

Following the ear incident, Vincent had a series of 'crises', to use his term. There have been many attempts to give these seizures of Van Gogh's a medical diagnosis, including articles in the prestigious *British Medical Journal* and the *Journal of the American Medical Association*. Although, hardly surprising, none of the literature considers the possibility of seizures like those of St. Teresa of Ávila. Like her, Vincent's attacks included dizziness and fainting.

Van Gogh seemed to sense on an intuitive level that there was a link between his suffering and being in the South of France: "I shall always be convinced that the South is responsible for this illness of mine." [95]

Why did Van Gogh's state spill over into such a profound crisis that he mutilated his ear and then took his own life only a year and a half later? There was obviously a great deal of psychological wounding in Vincent's family; his younger brother committed suicide and his sister spent nearly forty years in an asylum. The process of spiritual emergence, turning into emergency, as we have seen, brings all of that to the surface. What was unconscious starts to break through, exacerbating the kind of emotional pain that Vincent possibly fought so hard to keep at bay all his life.

Another major factor was that Vincent did not eat properly. That might seem neither here nor there, but if we remember that the ascetics of centuries gone by starved themselves in order to heighten their spiritual experience, then we start to get close to the effect of drinking 23 cups of coffee and eating only a little bread over several days. Part of the reason was that Vincent spent nearly all the money Theo gave him on paints, canvas and other supplies. Added to a tendency towards what Emile Bernard, his close friend and Christian mystic, called almost 'biblical mortifications'. Throughout his life Van Gogh purposefully denied himself; going out in winter without an overcoat, eating very small meals even when plenty was available. Whilst this was his way of aspiring to follow Christ's example, from another perspective it might look like self-harm, a coping strategy to manage emotional and psychological pain. Mutilating his ear and taking his life could be seen as further manifestations of self-harm, further along the spectrum.

Eating is one of the most grounding things we can do when in spiritual crisis. Keeping regular meals going is one of the fundamentals. Whatever the reasons for Vincent hardly eating, coupled with spending weeks at a time painting furiously under the energy of the burning sun and challenging material erupting from his unconscious, we can see that these conditions could easily precipitate such a severe spiritual crisis.

Van Gogh's suffering and spiritual emergency came to an end when he shot himself in the stomach. His death came at a time when Theo was experiencing financial difficulties. It is worth remembering Vincent's views on death. Years earlier, on hearing of the death of a favorite cousin, Vincent commented that it left him 'sorrowful yet always rejoicing'. This is the view of someone who has a solid faith in life after death, in the certainty of union with God, the Source or whatever language you prefer.

His spiritual take on death, as opposed to a purely material, physical perspective, also came across when he talked about the symbolism in his painting of a reaper:

> ...I see in him the image of death, in the sense that humanity might be the wheat he is reaping... But there is nothing sad in this death, it goes its way in broad daylight with a sun flooding everything with a light of pure gold. [96]

And we know from his other letters that he used the sun, light and gold to symbolize God and the Divine. Elsewhere he talked of death as being no more than the falling of a leaf from a tree. Without denying the huge distress that caused him to shoot himself, it seems Van Gogh had a strong sense of something after death, of the 'infinite' he longed for.

Tragically, some people do commit suicide when going through spiritual emergency. The intensity of the suffering becomes too much for them to bear. Reaching out for the right support is vital. We will look at this in Part III.

CARL JUNG

When I first came across Jung's autobiography, *Memories, Dreams, Reflections*, I was fascinated to read about the near-death experience he had at the age of 69, during a heart attack. The difficulty he had in adjusting back to everyday life after his encounter with the other realm struck me as the stuff of spiritual crisis. In fact, as we have seen, near-death experiences can trigger spiritual crisis.

I read that book at a time when his other major autobiographical piece, his *Red Book*, was not available. Jung's family kept it under wraps and it was only finally published 50 years after his death. It caused quite a stir when it came out in 2009 and it is easy to see why his family resisted publication for so long. To my mind, it is one person's journal of their spiritual emergence and emergency. What Jung shows us in the *Red Book* is the possibility of navigating such treacherous waters safely. He does this by example, through containing, integrating and understanding his awakening of consciousness.

Born in 1875, the Swiss psychiatrist was brought up in a rural parsonage. He studied medicine and psychiatry, putting spirituality at the heart of psychology in his work. He taught and had a private therapy practice. He gave birth, in effect, to psychotherapy as we know it today by taking it from its then focus on treating neurosis to addressing personal and spiritual development. He wrote:

> The main interest of my work is not concerned with the treatment of neurosis but rather with the approach to the numinous... [which] is the real therapy. [97]

Jung is known for his groundbreaking ideas on archetypes, the personal and the collective unconscious, the role of the anima and animus (Latin for 'soul'), the process of individuation and the psychology of spiritual experience. He published profusely on these subjects. It is perhaps less well known that all of these major theories, indeed his life's work, stemmed from the period of crisis he went through during his early forties, which resulted in the *Red Book*.

The years, of which I have spoken to you, when I pursued the inner images were the most important time in my life... the numinous beginning, which contained everything, was then.[98]

His period of intense introspection, of descent into inner chaos, began in his 40th year, at a time when he felt he had achieved everything he wanted on the material level. He had a flourishing career, success, a large family. Then in the autumn of 1913 Jung had horrific, apocalyptic visions of Europe flooded, with the floodwaters turning to blood and 'the drowned bodies of uncounted thousands'.

It was these visions that prompted him to start exploring his interior world, fearing as he did that he was 'menaced by a psychosis'. Jung kept a detailed record of this in-depth exploration of his psyche. It was not until World War I broke out that he realized, far from being personal, the visions related to Europe's fate. It was only then that he started to compile his famous *Red Book*, from the notes he had been keeping.

It was this experience that led to his theory on the collective unconscious, as he came to realize that he had confused material from the collective with his personal unconscious. This commonly happens to those going through spiritual emergency. As Jung acknowledged, it is not easy to distinguish between the two.

By the start of the war, however, the inner process had taken on a life of its own and he carried on working on the *Red Book* on and off until 1928. It is arguably the most important book he ever worked on, yet it remained unpublished during his lifetime, concerned as he was that it would be misconstrued, that in places it would read, to the undiscerning, like crazy ranting.

Man's task, man's destiny, Jung tells us, is to "create more and more consciousness" [99]. His *Red Book* is one man's journey in the awakening of consciousness, the fears and dangers, the trials and tribulations of surrendering to the process of spiritual emergence, the process he came to call 'individuation'. From reading the *Red Book* we discover that he too plumbed the depths of so many of the issues that become of concern during spiritual emergency. He too felt the fear of going mad, the fear of dying, the tension of opposites, the importance and power of the irrational, of making space for that which does not make sense to the logical, rational mind. Jung explored the nature of the soul, of divine madness, and so much more. His *Red Book* is one of the most important records we have of the journey through spiritual emergency. So much of what Jung wrote will resonate with those who have personal experience of it.

What particularly distinguishes the *Red Book* and makes it such an extraordinary account was Jung's ability to contain the experience, to hold it himself. By this I mean his ability to observe what was happening, maintaining just enough distance from it, just enough space around it, to not be totally overwhelmed. Like many therapists, who are trained to notice what is happening in the present moment without reacting to it, he was in fact using the skill of mindfulness, without calling it that. Most people find they cannot function at the everyday level when flooded with material from the personal and collective unconscious, to use Jung's terms. Jung, remarkably, was able to do so. Even so he had to make certain adjustments. He was forced to abandon his academic career.

> My experience and experiments with the unconscious had brought my intellectual activity to a standstill… I found myself utterly incapable of reading a scientific book. This went on for three years.[100]

This is in line with what we would expect, as rational, left-brain functions become difficult, if not impossible, during spiritual emergency. He was, however, able to carry on seeing clients and this makes perfect sense too, immersed as he was in the worlds of symbolism and mythology. In fact, we could expect him to have been a particularly insightful and intuitive therapist during that time. Between 1913 and 1914 he had an average of six clients a day, five days a week. Jung himself said that it was only carrying on working and the knowledge that "I have a wife and five children, I live at 228 Seestrasse in Küsnacht"[101] which assured him that the material world, and himself in it, was real and existed.

Jung's journeying, a sort of Jungian version of Shamanic journeying, took the form of active imagination, of waking fantasies that he consciously and deliberately entered into. Along the way he has fantastical and gruesome adventures and many conversations with the figures he encounters. Early on, he finds himself in the desert, the loneliest of places, as indeed spiritual emergency can be. Unsure, he asks:

> My soul, what am I to do here? But my soul spoke to me and said, 'Wait'. I heard the cruel word. Torment belongs to the desert … Nobody can spare themselves the waiting and most will be unable to bear this torment…[102]

Jung knows only too well the torture of the agonising wait, sitting with the unknown, facing the unknown, all the time with the fear that he is losing his mind. And yet all he can do is sit with it, bear with it. The greater his ability to bear with it, the better he will fare. Likewise, all we can do is sit with it, bear with it. The greater our ability to bear with it, the better we will fare.

As the process takes hold he feels himself descending into chaos:

And suddenly to your shivering horror it becomes clear to you that you have fallen into the boundless, the abyss, the inanity of eternal chaos. It rushes towards you as if carried by the roaring wings of a storm, the hurtling waves of the sea.[103]

Jung has to deal with fear, fear of losing his mind, "the fear of death, which spread like poison everywhere in my body"[104], and just plain fear:

You dread the depths; it should horrify you, since the way of what is to come leads through it. You must endure the temptation of fear and doubt and at the same time acknowledge to the bone that your fear is justified and your doubt is reasonable...

I have had to recognize that I must submit to what I fear; yes, even more, that I must even love what horrifies me.[105]

What Jung is recognizing here is our need to step towards those aspects of our experience, of the crisis, that we instinctively want to push away. By resisting, we only exacerbate matters. Likewise, his soul openly encourages him to let madness in:

My soul spoke to me in a whisper, urgently and alarmingly:... You wanted to accept everything. So accept madness too. Let the light of your madness shine, and it will suddenly dawn on you. Madness is not to be despised and not to be feared, but instead you should give it life.[106]

Of the various figures Jung encounters during his journeying, one is a fellow psychiatrist. Jung describes how the 'fat professor' takes one look at him and the book he has with him (*Imitation of Christ* by Thomas à Kempis) and declares a case of religious paranoia. When Jung protests that he feels 'perfectly well', the fat professor replies: "Look, my dear. You don't have any insight into your illness yet. The prognosis is naturally pretty bad, with at best limited recovery."[107]

It was precisely because of this, because society could not accommodate the irrationality of divine madness, of spiritual emergency, because Jung feared the response of those in his field and beyond, that he was wary of publishing this innermost journey of his, the *Red Book*.

In the text Jung alternates the fantasies themselves with commentary. It is here that he gives an understanding, newly gained, of what his soul, a soul, is — an aspect of f that has an independent life and is an independent being in its own right. And ot always happy with his soul's take on things: "[M]y soul spoke to me saying,

'My path is light.' Yet I indignantly answered, 'Do you call light what we men call the worst darkness?" [108]

Here, in frustration, Jung sums up the paradox of spiritual emergency, that what appears to be our worst nightmare can, in fact, be our greatest blessing, that there is light in the darkness. Jung's tone is one of disbelief; it can be so very difficult to know this truth when we are in the thick of it all.

Ultimately Jung surrenders to the process, which is all any of us can, and indeed must, do:

> I lock the past with one key, with the other I open the future. This takes place through my transformation. The miracle of transformation commands. I am its servant … [109]

Jung understood divine madness. He recognized the vital place of the irrational, the paradoxical, the inexplicable. He totally grasped that the way to make sense of spiritual emergence and emergency is through the symbolic and the mythological. Psychiatrists who followed his lead and paid attention to the mythic realm, like the American John Weir Perry, also achieved results.

There is no doubt that Jung saw the awakening of consciousness as the single most important thing we can dedicate ourselves to. He showed us that it is possible to teeter on the brink of so-called madness, to find healing and growth, awareness and insight in that place, to come back to the everyday world and to live extraordinary lives at a higher level of awakening and consciousness.

In this chapter we have been looking at the relationship between spiritual emergency and creativity, through the lives of three remarkable creative geniuses. Several of the people whose stories we heard in Part I are very good examples of the enhanced creativity that can come following spiritual emergency. Having been committed to hospital (sectioned) three times, as Annabel opened up fully to her spiritual life and the healing and growth that came with that, she embarked upon a two-year art course. She followed that with a writing course and then started offering workshops using various creative media to help participants explore their inner life. Emma took her documentary film making to a whole new level and Kimberley started using her artistic gifts to offer healing to others, through intuitive energy paintings.

The Present Day

In this final chapter exploring spiritual emergency through the ages, we look at four contemporary figures, some more well-known than others. These are people who have been through a crisis of spiritual awakening, or spiritual emergency. Some have managed to move through it, others have not.

MARIE MOORE

Please take care when reading this section. You may find it disturbing. If you are feeling very sensitive or vulnerable right now, make sure you get some support, talk to a friend or come back to reading this section when it feels okay to do so.

On a Sunday afternoon in April, 2009, at a firing range in Florida, Marie Moore shot her son at point blank range in the back of the head and then turned the gun on herself. Her son, Mitchell, died instantly; she died not long after in hospital.

This horrific event made the news internationally, with questions being raised about why Marie had been allowed into the firing range given her previous psychiatric history. She had been in hospital seven years earlier, held under the Baker Act, the equivalent of being sectioned in the UK. Her ex-husband, Mitchell's father, said she had previously tried to commit suicide at the same firing range in Casselberry, near Orlando.

Suicide notes apparently led police to a locker containing 67 journals, outpourings of her mental anguish, which most of us cannot even begin to imagine. Looking at what she said in her suicide notes and audio tapes we see, in the language she used, the tell-tale signs of spiritual emergency. She signed herself 'failed Queen', believing, according to one report, that God had made her a queen and she had failed. The archetype of the queen is one that comes up during spiritual emergency, although not as often for women maybe as the mother of Christ, Mary. It is typical of the ego inflation covered earlier in the book. The sense of having failed God also comes up repeatedly in spiritual crisis. Marie felt she was the anti-Christ, referring to this more than once.

This is not to deny that Marie obviously needed medical care and attention, that her mental torment was seriously deranging her thinking. As we have seen, spiritual emergency and mental health issues are not mutually exclusive. They very often go hand in hand. Psychiatric staff with a transpersonal training would hopefully have been able to help Marie more than mainstream services had.

The most telling aspect of Marie's experience, in terms of spiritual emergency, was covered in the UK newspaper, the *Daily Mail*. According to the article, police reports revealed that in her audiotapes Marie spoke of:

> the mental 'misery and torment' she suffered amid hallucinations that had her convinced at times that she was being buried alive, eaten by ants, burned at the stake and gassed.

This is the stuff of spiritual emergency. This is material exploding from the personal and collective unconscious, the two inextricably intertwined. Some of it may be past-life related, such as being burnt at the stake, which comes up repeatedly as a theme in past-life regression work, as does being gassed. Some of it may have been from the collective unconscious, not strictly Marie's material at all.

One report, in the UK newspaper *The Telegraph*, mentioned that Marie thought the world was coming to an end. This kind of apocalyptic fear, along with the vivid scenes of her own death, is highly typical of spiritual emergency. Such concerns mark the disintegration of the psyche. When spiritual emergency is properly supported, Perry argued, this is in the service of its re-integration into a healthier whole, functioning at a higher level. Tragically, for Marie, she did not get the help she needed to be able to heal the wounds she was carrying.

As we move further into the Mayan End of Time, described by some commentators as covering the years 2007-2015, more and more people may start having apocalyptic visions of death and destruction. Jung was by no means the only one to have visions of the horrific bloodshed of World War I. Other creative people tuned in to the coming events; artists painted violent, bloody scenes. As Jung found, it is when we make the mistake of taking such material from the collective unconscious to be personal that it becomes threatening to us individually.

Whatever we make of the disturbing visions Marie experienced, for somebody with a fragile psyche and ego wounding, it is easy to see how such material could be completely overwhelming. The carnage happened at a time of year when the archetypal themes of death and rebirth, or resurrection, are more intensely alive than at any other time. It took place on Sunday 5 April. The following Friday was Easter Friday, in other words the most potent time of year in the Christian calendar. The title of the *Daily Mail* article read:

I had to send my son to heaven and myself to hell.

To conclude that Marie was going through spiritual emergency is not to say that she was not in desperate need of mental health care. Nor is it to say that in such an extreme case there are any easy answers. But a physician who dismisses as irrelevant the content of the nightmare scenes that plagued her would not have been the kind of health care she needed.

Until consensual reality broadens to include an understanding of the archetypal, of the collective unconscious, of past-life experiences, then tragically people in Marie's situation will not get the care they need. Broadening the consensual reality of mainstream psychiatry is quite literally a question of saving lives.

ECKHART TOLLE

Eckhart Tolle's name has become synonymous with 'presence' and the 'power of now', the key tenets of his teaching. Born in Germany, he became a researcher at Cambridge University, in the UK, and experienced a major awakening in his 30th year. His first book, *The Power of Now*, became an international bestseller, translated into 33 languages, followed by videos, CDs and DVDs. His second book, *A New Earth*, was featured by Oprah Winfrey when she interviewed Tolle for a web seminar that had 700,000 viewers over ten weeks. He is now teaching through the medium of web TV. Even watching Tolle online, he has a palpable presence, a sublime stillness descends. His very being is stilling, bringing one into the present moment. His cheeky chuckle is delightful.

The profound awakening that Eckhart Tolle experienced was accompanied by an equally profound crisis or 'dark night of the soul'. Already during his childhood Tolle was unhappy and thought about how to kill himself. When he was 13 he went to live with his father in Spain. His father was unconventional and when Tolle said he did not want to go to school, his father took him at his word. So Tolle followed his own interests, including astronomy, and went to language classes. Someone visiting the family left behind some books by a spiritual teacher on awakening. When Tolle read them something stirred within him. During his teens and 20s his mental suffering continued. 'I lived in a state of almost continuous anxiety interspersed with periods of suicidal depression.'[110]

He worked as a tour guide and language teacher, moving to England at the age of 19. There he started to follow a more academic route in search of some answers. Despite the success of his university career, his life still felt meaningless to him. Looking back, he says his professional life was based on fear at the time. Then one night, not long after his 29th birthday he reached a crisis point. His sense of dread and despair plumbed new depths.

> One night I woke up in a state of dread and intense fear, more intense than I had ever experienced before. Life seemed meaningless, barren, hostile. It became so unbearable that suddenly the thought came into my mind, "I cannot live with myself any longer." The thought kept repeating itself several times. Suddenly, I stepped back from the thought, and looked at it, as it were, and I became aware of the strangeness of that thought: "If I cannot live with myself, there must be two of me — the I and the self that I cannot live with." And the question arose, "Who is the 'I' and who is the self that I cannot live with?" There was no answer to that question, and all thinking stopped. For a moment, there was complete inner silence. Suddenly I felt myself drawn into a whirlpool or a vortex of energy. I was gripped by an intense fear, and my body started to shake. I heard the words, "Resist nothing," as if spoken inside my chest. I could feel myself being sucked into a void. Suddenly, all fear disappeared, and I let myself fall into that void. I have no recollection of what happened after that. The next morning I awoke as if I had just been born into this world. Everything seemed fresh and pristine and intensely alive. A vibrant stillness filled my entire being. [111]

This extraordinary experience then took years to integrate. For the first two years Tolle sat on park benches. All the trappings of the 'self' fell away. He describes this in the introduction to *The Power of Now*:

> A time came when, for a while, I was left with nothing on the physical plane. I had no relationships, no job, no home, no socially defined identity. I spent almost two years sitting on park benches in a state of the most intense joy. [112]

Many who have been through spiritual emergency will recognize this falling away of everything that is familiar. People can find themselves jobless and homeless and undoubtedly some have spent years sitting on park benches or the equivalent, not necessarily in a permanent state of bliss, but because they find themselves unable to function at an everyday level. The difference for Tolle, as he writes in *The Power of Now*, was that he could function in the world.

Another aspect of Tolle's experience that will be familiar to those who have been through spiritual emergency is how very long it can take to integrate such a powerful transformation, perhaps even more so where the process has not unfolded as smoothly as it seems to have done for Tolle.

Some will also recognize the physical shaking that Tolle speaks of, which seems to very frequently accompany these extraordinary happenings. It can range from a gentle sense of the body vibrating to mild shaking or much stronger, even violent shaking. There are several potential reasons for this. Our individual vibration of energy may be raised, we may be releasing trauma held in the body, or we may be shaking in terror.

What Tolle has to say, in an interview with Sounds True, about the relationship between suffering and awakening is very relevant to our theme of spiritual emergency.

> "For most people, spiritual awakening is a gradual process. Rarely does it happen all at once. When it does, though, it is usually brought about by intense suffering. That was certainly true in my case."

The gentle, gradual process of awakening that many go through does not tend to give rise to spiritual crisis, whereas when the process speeds up it very often does. Not only can awakening be triggered by deep suffering, but the process itself can bring with it a level of suffering never previously experienced. Many of us go through a partial awakening as opposed to Tolle's more complete awakening. The Sounds True interview continues on this theme:

> ST: In your own life story there seems to have been a relationship between intense personal suffering and a breakthrough spiritual experience. Do you believe that for all people there is some connection between personal suffering and the intensity that is needed for a spiritual breakthrough?

> ET: Yes, that seems to be true in most cases. When you are trapped in a nightmare, your motivation to awaken will be so much greater than that of someone caught up in a relatively pleasant dream. On all levels, evolution occurs in response to a crisis situation, not infrequently a life-threatening one, when the old structures, inner or outer, are breaking down or are not working anymore. On a personal level, this often means the experience of loss of one kind or another: the death of a loved one, the end of a close relationship, loss of possessions, your home, status, or a breakdown of the external structures of your life that provided a sense

of security. For many people, illness — loss of health — represents the crisis situation that triggers an awakening. With serious illness comes awareness of your own mortality, the greatest loss of all.

And the same principles apply on a global level. During the web seminar with Oprah Winfrey, Tolle refers to this. A viewer asked why this was happening now. Why were some 700,000 watching a show encouraging an awakening of spiritual consciousness? Tolle answered:

> "It's happening now because we're reaching a crisis point. Very essential things don't happen until there's an absolute need for them to happen. If you look at the history of the 20th century, that gives you a taste of what it will be if there is no major shift."

Conservative estimates conclude that more than 100 million humans were killed by other humans in the 20th Century, Tolle said.

> "It's unbelievable insanity when you look at that history. And so, if there's no shift in consciousness, we will go downhill very quickly, because we're already in the process of destroying the planet. But there will also be continuous conflict, collective conflict, and eventually, then, humanity would collapse."

It is this crisis point that we have reached collectively which is triggering a world crisis of spiritual awakening. We have reached a state of global spiritual emergency and many of the same principles apply as at the individual level. It is a time of great opportunity as well as potentially a time of great danger. To repeat Tolle's words: 'On all levels, evolution occurs in response to a crisis situation.'

DAVID SHAYLER

David Shayler, the former MI5 agent, hit the front pages in 1997 when he leaked shocking information to the UK newspaper the *Mail on Sunday* about the intelligence services investigating Government Ministers. He fled the UK, returning voluntarily three years later, when he was tried and imprisoned for breach of the Official Secrets Act. The next time Shayler hit the media limelight was in 2007, when journalists, to their glee, discovered him living in a squat, cross dressing and claiming to be the next Messiah.

Inevitably, the media's take was that he was mentally ill, although his former girl-friend, Annie Machon, was unwittingly nearer the mark when she described him as having had 'some form of severe breakdown'. As we know, breakdown can lead to breakthrough. For those familiar with the course of spiritual emergence and emergency, it does not take long to realize that Shayler was going through that very process. Let us look at some of the background first.

Shayler had a habit of publishing unpalatable truths. At Dundee University, as editor of the student newspaper *Annasach*, he was responsible for printing extracts of the then banned book *Spycatcher*. Despite the Thatcher government's attempts to silence the author, former MI5 agent Peter Wright, or perhaps because of those attempts, the book sold over two million copies. After graduating with an English degree, Shayler worked for the *Sunday Times* for six months. Some 18 months later, looking through job adverts in *The Observer*, he spotted what he thought was a media position. Applicants needed to have an interest in current affairs, common sense and an ability to write. The job turned out to be for the secret services. Shayler's penchant for publishing government secrets as a student seems to have passed his MI5 recruiters by, unless they thought it provided the perfect cover. He worked there for six years, leaving in 1996.

The following year, in August, he passed documents to the *Mail on Sunday* showing that MI5 had previously investigated Labour Party Government Ministers Harriet Harman, Peter Mandelson and Jack Straw. The day before publication he fled the country. When the British Government attempted to extradite him the French courts refused, on the grounds that the request was politically motivated. In the meantime he had spent four months in a French prison without charge. He returned to the UK voluntarily in August 2000 and was tried and convicted. Shayler was given a six-month prison sentence, serving only eight weeks of it, because the four months on remand in France were taken into consideration.

Watching newsreel and reading interviews of Shayler, all the elements of spiritual emergency are clearly there. He describes, in one YouTube video, the impact of being imprisoned for telling the truth, the dark place that took him to and how that led to the beginning of his spiritual journey. Dark places, traumas of any kind, are renowned for triggering spiritual crisis. St. John of the Cross' wrote his poem *Dark Night of the Soul* while he was in prison.

Unfortunately Shayler seems to have got 'stuck' mid-process, convinced that he is the latest reincarnation of Jesus. It is very common when the energy of the Christ Consciousness enters an individual for the lower self and higher Self to become confused and for the person to feel they are Mary or Christ or, as in Shayler's case, a reincarnation of Christ. Usually, however, the person moves through this, whereas Shayler, certainly at the time he was last interviewed by Channel 4 News, had not.

In that interview, dressed all in white, looking slimmer, healthier and more at peace than in any of the earlier media shots, Shayler says:

> "What I'd say to people is: Do I look mentally ill? Do I sound mentally ill?…The reason I'm putting out this message is because I am absolutely convinced — as convinced as I can be of anything in this world — that the Universe is changing shape and humanity has to prepare for that, and that I am here to help teach people."

Whilst this might sound an odd statement to the mainstream media, it is of course precisely what the likes of Eckhart Tolle are saying, that there is a huge shift in human consciousness taking place. Tolle sold over five million copies of his book *A New Earth* on the strength of it. Other comments of Shayler's will resonate strongly with those who have been through spiritual crisis:

> "Suddenly my whole life made sense, I felt a sense of peace, I suddenly realized why it had been how it had…Why I seem to get such a strange deal from the Universe, when I seem to be trying to tell the truth about everything."

There was obviously a particular moment, on 29 June 2007, when Shayler had a very powerful spiritual experience. Speaking of that, he told Cahal Milmo of *The Independent*, the UK newspaper:

> "It was a deeply, deeply humbling experience. I felt this incredible energy, way beyond any sexual or physical experience. It was way, way beyond where I had been before. What do you do in those circumstances? I fell to my knees and prostrated myself. I had become the spirit of Jesus."

While it seems that Shayler had a genuine experience of the Christ Consciousness, where he was at odds with mainstream media was in seeing himself as the reincarnation of Jesus. This is, however, both typical and common of someone going through spiritual crisis. St. Teresa of Ávila describes how she fell to her knees when she felt the energy of the Christ Consciousness enter her. The difference between her and Shayler is that St. Teresa had a whole context, a very strong tradition, within which to understand and integrate her experience. To remind ourselves of what Assagioli, says of such experiences:

...the inflowing spiritual energies may have the unfortunate effect of feeding and inflating the personal ego ... instances of such confusion ... are not uncommon among people dazzled by contact with truths which are too powerful for their mental capacities to grasp and assimilate.

Receiving the energy of the Christ Consciousness, which seems to be what happened to Shayler, is a completely extraordinary, unfathomable experience. It is hardly surprising that it is beyond the mind's grasp, that the ego appropriates it to itself, in its dazzlement and confusion.

Understood within the context of spiritual emergence, what Shayler had to say about his cross-dressing and his transvestite persona, Dolores, also makes sense:

"Obviously I'm not living full-time as Dolores. Transvestism is not the same as trans-sexuality and I'm perfectly happy as a bloke. I probably dress once or twice a week at the moment. And no, I'm not gay. And no, I haven't had a breakdown — I've been dressing as Dolores — not Delores — for years. In fact, my ex-girlfriend Annie initially encouraged my exploration of my feminine side — although only in private... Any spiritual teacher will tell you that the ultimate goal of the journey is to combine and balance the masculine and feminine. ...If you are offended by or condemning of a man in a frock then you are not on the journey of love."[113]

On the same subject, he told *The Independent*:

"A bloke in a frock is whole lot less offensive than blowing up innocent people in Iraq and Afghanistan."

It is important not to romanticize what Shayler has been through. Much of what Shayler said is very typical of a person going through spiritual emergency who has not yet come out the other side. Seen in that light, a great deal of it makes an awful lot of sense. Having his experience pathologized so widely by the media and those around him, would only contribute to his sense of being the reincarnation of Christ remaining entrenched, along with some of his other unusual convictions. We can only hope that in the time that has elapsed since his last interview he has received the support he needs to fully integrate his spiritual emergence and emergency, to go on to share what he has to offer at this time of huge transition.

AMMA

Today Amma, as she is affectionately called, travels the world, hugging and offering her blessing to thousands in every major city she visits. As well as her gruelling schedule, Amma directs impressive charitable works, overseas building projects of hospitals, temples and homes for the poor, and much more. There was a time, however, when she was considered insane. Her process of spiritual emergence, of becoming the Realized being that she is today, took her through the deep waters of spiritual emergency.

Born in 1953 in a small fishing village in Kerala, Southern India, Amma's is a Cinderella story. Her birth was surrounded by many auspicious signs. The newborn, named Sudhamani, instinctively held a yoga and meditation posture, the lotus position, with crossed legs. Despite the powerful dreams of Hindu Gods and Goddesses both parents had had during her mother's pregnancy, they failed to recognize the providential signs. Believing the strange posture to be some horrible bone disease they sought the opinion of doctors. The baby's skin was also a strange shade of dark blue, which lasted over six months. Eventually it turned dark brown, but this did not endear the baby to her mother, as her skin tone was so much darker than the rest of the family's.

From the age of about two, Sudhamani was already making up and singing devotional songs to Krishna. While playing, she would sometimes become withdrawn and drift off, transported to another world. These absorbed states gradually intensified as she grew older and turned into spontaneous altered states of consciousness. This behavior caused her family to think of her as distinctly odd and possibly suffering from a psychological disorder. For various reasons, Sudhamani was treated as the family servant from very early on. Her strange ways, the darkness of her skin, the fact that she was the eldest daughter still at home and her mother's poor health, having had 13 pregnancies with eight surviving children, all contributed. With never ending household chores from before dawn to long after dusk each day, her mother and eldest brother played the roles of the Ugly Sisters, heaping abuse and cruelty on her.

Sudhamani started school, but struggled because of all the housework. She was often late, although a very bright student with an extraordinary memory. After a few years, when she was ten, her parents decided to take her out of school, to concentrate on the cleaning, cooking, clothes washing, looking after the younger children and the livestock. She would carry out all her tasks praying, singing all the while to Krishna. When scolded she took to repeating the Lord's name and sacred chants under her breath. Seeing her lips constantly moving while she was working, one of her brothers would taunt her, saying it was a sign of madness. At night time she would go to the family shrine room to pray and meditate, to sing and dance to her Lord, as this was the only time available after all her work was finished. It left little time for sleeping. If she did oversleep in the morning, her mother would sometimes pour a jug of cold water over her.

She also punished Sudhamani harshly when she caught her taking food. Sudhamani never let on that she was giving it to some of the villagers who were living in desperate poverty, in danger of starvation. When she stole jewelry from a relative's home for the same reason, she was seen as bringing dishonor on the family and was beaten even more severely. Attempts to marry Sudhamani off met with total failure as she made sure she behaved so appallingly that suitors quite literally fled the house.

The girl's absorbed states of altered consciousness often meant she lost consciousness altogether, much like some of the mystics we looked at earlier. Family members would frequently have to go off in search of her when she failed to return from some errand or other. At times, when her father found her unconscious lying on the sand by the shore, he would have to carry her home, having failed to revive her.

Such deep states sometimes put Sudhamani in danger. The family latrine was perched over the backwaters and on occasion she would fall in. Another time she was in a little rowing boat crossing over to the other side of the waterway. As she drifted off, transported by the sound of the lapping waves, she was oblivious to the large boat heading straight towards her, despite the loud noise of its engine and the frantic shouting of people on the shore. She came to just in time to steer out of the way.

Deeply devout from such a young age, Sudhamani longed for Krishna, to see Krishna, to be one with Krishna. Her experience of worldly life was one of pain and suffering, which made her see the mundane world as a selfish place, where there was no truly selfless love. After years of praying and meditating, of devotional singing and ecstatic dancing in states of bliss, after years of yearning for Krishna, of searching for him, of invoking him, finally her Prince, her Lord came. Looking back, she recalls:

> I used to look at Nature and see everything as Krishna. I could not even pluck a single flower, because I knew that it was also Krishna. When the breeze touched my body, I felt that it was Krishna caressing me. I was afraid to walk because I thought, 'Oh, I am stepping on Krishna!' Every particle of sand was Krishna for me. Now and then I strongly perceived myself to be Krishna. Gradually this became a natural state. I could no longer find any difference between myself and Krishna... One day I strongly felt the urge to be absorbed in the Supreme Being without returning. Then I heard a voice from within saying, 'Thousands and thousands of people in the world are steeped in misery. I have much for you to do, you who are one with Me.'[114]

Soon after, the local villagers were to witness the outward manifestation of the Krishna Consciousness she had fully embodied. At a devotional gathering one day, she went into an altered state and her features took on the appearance of Krishna, as he is traditionally portrayed. Her skin turned the strange shade of dark blue that her parents had seen at birth. Those present felt that Krishna himself had come to bless them, but some wanted proof and pressed her for a miracle. At first resisting, she asked a man to fetch a pitcher of water. Asking him to dip his fingers in, when he drew them out they were dripping milk. She asked another skeptic to dip his fingers in and this time they were covered in *panchamritam*, a sweet Indian pudding made from milk, bananas, sugar and raisins. The jug was passed around for all to taste for themselves. Word spread quickly and many came running to see. Over a thousand people tasted the *panchamritam* and still the jug did not run dry.

From then on Sudhamani publicly manifested her Krishna consciousness three times a week and more and more devotees came. She was 22 years old. Her family seemed to veer between believing she had schizophrenia and thinking she was temporarily possessed by Krishna, when they were respectful and in awe. The rest of the time, however, they still treated her as a servant and were still abusive towards her. Of that period she says:

> I was able to know everything concerning everyone. I was fully conscious that I, myself, was Krishna, not only during that particular time of manifestation, but at all other times as well. I did not feel, 'I am great'. When I saw people and knew their sufferings, I felt immense pity for them. I was conscious of devotees offering salutations to me and addressing me as 'Lord'. I could understand the sorrows of the devotees, even without being told.[115]

Not everyone was a devotee of Sudhamani. Locally those antagonistic towards her organized themselves into the Committee to Stop Blind Beliefs. They did their best to make life difficult for her and for those visiting. There were at least two attempts on her life. Her eldest brother and a cousin were involved in one such incident. Not long after, this brother, who terrorized the whole family, banned Sudhamani from the house. She had disobeyed him and carried on with the public audiences, bringing, in his opinion, shame upon the family. Sudhamani was left to live outside in the courtyard.

This period from 1977 was the deepest time of her spiritual progression. One day sitting deep in meditation with her eyes open, she saw the most stunning orb of light, the color of the setting sun. Out of it emerged the figure of a Goddess wearing a beautiful crown. Until then Sudhamani had only worshipped Krishna, as the highest divinity. This was the beginning of her awakening to the Divine Mother, Devi. From then on she sought out Devi, calling on her and invoking her as she had previously Krishna.

Whereas before she had taken care not only of herself, but the rest of the family too, Sudhamani now became unable to look after herself. Living like a wild creature, sleeping outside on the ground, she stopped washing and eating. She drank milk direct from the udders of a cow and an eagle regularly brought her fish, which it dropped into her lap. Sometimes her mother would cook it, but more often than not Sudhamani ate it raw. Some neighbours would take pity on her. Finding her in one of her altered states, unconscious, lying in the mud, the women would wash her, dress her in clean clothes and feed her.

During this period of spiritual emergency, Sudhamani experienced the terrific heat in her body typical of the rising Kundalini energy. Feeling as if her body was burning, she would go and stand in the backwaters, meditating with water up to her chin. She avoided all human contact and dug big holes in the earth in order to hide. Everybody considered her to have finally succumbed to her mental illness.

She came through, however, at the end of this time experiencing the Divine Mother merging with her in a state of supreme bliss. Following this mystical union she came to be known as Mata Amritandamayi, Mother of Divine Bliss, or Amma, meaning 'Mother'. And she started manifesting the Devi consciousness publicly, as she had the Krishna consciousness.

As she travels around the world, she is known as the 'hugging saint' because she gives each and every visitor or devotee a personal hug, receiving thousands of people for hour after hour, over two or three days in each city. This simple yet profound practice came from her younger years when seeing someone, maybe a neighbor or another villager suffering, she would instinctively offer them a warm hug. Seeing this blessing others wanted one too and it went from there to the extraordinary global *darshan* (blessed audience) that it is today.

Amma's is an example of the process of spiritual emergency left to unfold according to its own pattern. Her natural affinity with animals and the kindness of neighbors meant that her most basic needs were met. It is arguably because the process was not interfered with in any way, for example by mental health services, that she was able to break through to such a high level of Realization. Yet how many of us would have been able to trust the process sufficiently to leave her living in such conditions? How many of us would have been able to trust that she would come through, and not only come through, but as an Enlightened Being? This truly is the process of spiritual emergence and emergency taken to its extreme. It is all the more remarkable for the fact that it has taken place in our times.

Moving Successfully Through Spiritual Emergency: The Three Key Phases

CHAPTER 8

Phase 1
Coping with the Crisis
— The Key Tool

There is a key tool that can help in dealing with the actual crisis itself. I have come across no other tool that is as powerful and effective or as simple. We can learn to use it ourselves, to manage our own crises, or we can help others to use it, if we are supporting them through theirs. This approach has been around for thousands of years. It has now entered mainstream health care and is being used widely in the UK National Health Service and in US health care, thanks initially to the work of Jon Kabat-Zinn. It is Mindfulness.

My deeply held conviction in the power of Mindfulness to help us cope originally came from my own personal experience of spiritual emergency in 2003 and 2006. This was then strongly reinforced for me when I had the opportunity of helping my brother. At the time he was on the verge of breakdown. Two or three days of input from me, of learning the basics of Mindfulness practice, was enough to turn his situation around into breakthrough, despite him having no previous experience of the approach. It was miraculous; an astonishing illustration of the potent simplicity of Mindfulness. No wonder Thich Nhat Hanh called his book *The Miracle of Mindfulness*.

My personal experiences do not stand in isolation. Mindfulness has now entered the mainstream and, thanks to the strong evidence base, is accepted as an effective tool for coping with many conditions, including cancer and HIV. What Mindfulness does not do is cure the condition or make it go away. What Mindfulness does do is change our relationship to it, in our case spiritual emergency. I firmly believe that the practice of Mindfulness can alleviate any suffering we are experiencing, whatever its cause. The Buddha obviously did too, as he taught Mindfulness as a path to enlightenment, the overcoming of all suffering. His teachings on the subject can be found in the Satipatthana Sutta (see Resources).

What this means is that if you already have a strong Mindfulness practice you can make very good use of it, if you find yourself going through spiritual crisis. If you are in crisis now, and have not previously learnt Mindfulness, it could be very helpful to find a therapist who has a daily Mindfulness practice. They will use their skill to help you stay in the present moment. And if you have already been through more than one period of spiritual emergency, then it is well worth considering learning Mindfulness, when things are steady, so that you have this effective tool to draw on.

Note, however, that when someone is in full-blown crisis it is not appropriate to learn Mindfulness at that moment. This is because one of the key ways we learn is through meditating. By its very nature meditation builds energy, the kind we are desperately trying to dissipate and ground during many crises. It also has the effect of 'opening us up', whereas in crisis we need to 'close down', especially the crown chakra, which tends to get blasted open. Basically, meditation is contra-indicated, to use medical jargon, when in crisis. The only practice I personally would recommend, depending on individual circumstances, is the Body Scan. This is a body awareness practice and as such can be beneficially grounding and embodying.

What is the Difference Between Meditation and Mindfulness?

In meditation we bring focused, concentrated attention to an aspect of our experience, usually the breath. It tends to be time limited and is what we do sitting on the cushion or chair, or lying down in the case of the Body Scan. We are learning, or practicing, moment by moment awareness of whatever arises in our experience as we meditate. We then aim to take that moment by moment awareness of our experience, as it is happening, through the whole day. This is the practice of Mindfulness, in other words taking the awareness that we cultivate during meditation with us, during all our activities and interactions.

Having tried meditation, Kate found Mindfulness far more helpful.

> "Although I was tempted to try medication at times, I wanted to feel the grief [of childlessness] and felt drawn to find ways to manage my thoughts and emotions, now they'd reached a level where a degree of management might be possible. My early experiences with meditation had been off-putting and when I tried again using Buddhist techniques, my hyper-sensitivity still sent me into altered states. Through mindfulness, however, I found something long sought — a spiritual practice that feels natural, simple, enjoyable and free of conceptualization. Putting the focus on being in the present, moment-by-moment, throughout the day suits me far better than the intensity of sitting meditation sessions."

Mindfulness is about being with our experience as it is, with no frills, no stories added on, no resistance, no grasping, just gentle receptivity. It is about being with our experience in the present moment, without dwelling on the past, without worrying about the future. Why on earth, you may well ask, would you want to be in the present moment during spiritual emergency, if what you are going through is so horrendous? You may well discover that if you can really be with it, whatever 'it' happens to be, that it is not as horrendous as you thought. You can begin to explore it, rather than resisting it.

There is another important aspect of Mindfulness, not always recognized, that comes in too. It is also about kindness, being gentle and tender with ourselves in our struggles, without judging, without being our own worst critic. We can only expect ourselves to be present with the hell of what we are going through if we are also prepared to be very, very kind with ourselves in that place. That kindness includes finding ways of comforting ourselves, which we will look at in the next chapter.

Ultimately, Mindfulness is about freedom and surrender. To be totally in the present moment is extraordinarily liberating.

How to Be Mindful

The way that we come into the present moment is through our bodies and our five senses. It is actually very simple, if we can only remember our intention to be aware. So much of the time we are either on autopilot or multi-tasking, or both! When you are totally in the present moment, experiencing sounds, smells, tastes as well as the other senses of sight and touch, then you cannot also be dwelling on the past or worrying about the future.

COMING INTO THE PRESENT

- Right now, what can you feel physically?
- What are you aware of as you read this?
- How does the book feel in your hands? Hard? Heavy?
- What are you sitting on?
- How does that feel underneath you?
- What clothes are you wearing?
- Can you feel the fabric on your skin? How is that?

Do you see how, while you are focusing on the physical sensations in this present moment, the mind cannot be anywhere else but right here, right now?

7 WAYS MINDFULNESS CAN HELP

I have identified seven ways in which Mindfulness can help us in coping specifically with spiritual crisis. Of these I consider grounding and working with the intense fear to be the most important for spiritual emergency.

MINDFULNESS – 7 WAYS OF HELPING

- Becoming embodied
- Grounding the energy
- Coping with the fear
- Coping with unusual or painful physical sensations
- Getting some space around the experience
- Slowing the process down
- Surrendering to the process

1 – Becoming Embodied

As I mentioned, the way into the present moment is through our bodies and the five senses. This is the first way in which Mindfulness can help us during spiritual crisis. It can help us to literally become embodied, to connect with our bodies, to inhabit our bodies. If you are spiraling off into an inner world, if you are being swamped by the contents of your unconscious, if archetypal and mythological images and themes are taking over, then being in your body and connected with your body, can be an enormous relief.

It can be an enormous relief, for example, to go for a walk and be in your body, feeling its movements, the muscles, joints moving, feeling the soles of your feet making contact with the ground, through your shoes or, in summer time, feeling the deliciousness of the grass under your bare feet. This brings us to the second way in which Mindfulness can help us to cope with spiritual emergency.

2 – Grounding the Energy

Mindfulness can help us to ground the vast inrush of universal energy that Assagioli talked about. Becoming embodied and grounding the energy are very closely related. The more connected with your body you are the easier it will be to ground the excess energy.

Grounding is absolutely crucial to coping with spiritual emergency. I repeat, grounding is absolutely vital in helping us to get through. There is a lot of talk about grounding in Mind, Body, Spirit circles, but there is an aspect that is sometimes overlooked. If you are being blasted with the incoming energy and are feeling completely spaced out, the most grounding activities that exist will not have the desired effect unless you apply Mindfulness to the activity. The reason I can say this with such confidence is that I remember vividly on one occasion planting out young lettuces from their seedling tray, up to my wrists in rich, dark soil, knelt on the ground and still feeling my crown chakra blasted open, feeling totally spaced out and unconnected with my body and the earth beneath me.

What I am saying is that gardening, planting, or digging the vegetable patch are not necessarily grounding of themselves. We can still be miles away with our thoughts. This is where Mindfulness comes in.

GROUNDING – MINDFUL DIGGING IN THE GARDEN

You can apply this approach to any grounding activity:
- If you are digging, where do you feel the pull in your body? Which muscles?
- How do the palms of your hands feel? Rough? Sore?
- How are your feet? Warm and dry? Or cold and damp?
- How does the earth feel beneath your feet? Soft and squidgy? Or firm underfoot?
- What does the soil smell like?
- What sensations can you feel against your skin? The soft drizzle of the rain? A gentle breeze or stronger wind? The warmth of the sun?

The mind will wander off, especially with the psyche's pressing concerns during spiritual emergency. We simply, gently and persistently, keep bringing the mind back to the physical, to the body, to the sensations.

Other potentially grounding activities are ordinary, everyday things like chopping vegetables or washing up. But again, without Mindfulness, it is very easy to go off into a reverie, to be miles away, rather than being here now, feeling the warm, soapy water on our hands. That is not to say that grounding activities alone cannot be at all effective, but the effect will be watered down if done without Mindfulness.

On that lettuce-planting occasion, I discovered by chance soon after, that driving could be a more grounding activity for me than gardening. Whilst it may not have been wise or advisable to drive a car that day, I discovered that it required a level of concentration and attention to the present moment and a level of application to the physical coordination of, say, changing gear, that was grounding. Nowadays I tend to find driving a fairly ungrounding activity because of the speed at which we move in a car. So what can work one time might not another. And what works for me might not work for you. Each of us is different.

FINDING THE RIGHT GROUNDING ACTIVITIES FOR YOU

Experiment with finding what really works for you. Try several different activities specifically with that aim in mind. If it is something you enjoy doing, there will be more incentive! Many of the suggested activities below are quite physical. If you are not well enough or strong enough for them, then visualizing yourself doing them is the next best thing.

Examples of different grounding activities
- Gardening, especially digging
- Chopping wood
- Kneading bread
- Exercising, e.g. swimming, jogging or, if you need something more gentle, long walks, especially in woods, where the trees are rooted in the earth
- Physical work, e.g. decorating, building a wall
- Cleaning
- Eating heavy foods, which take quite a bit of digesting. Even if you are vegetarian it might be worth considering temporarily eating meat.
  Drastic times call for drastic measures.
- Lying on the ground, feeling the connection with the earth along the length of your back. This can work indoors, but if the weather is good,
  then outdoors is better. The alternative is to lean up against a tree.
- Lie on the floor on your back and get someone to hold your feet. Feel or imagine the energy being pulled down through your body.

If you notice that repeated efforts to get grounded, including practicing Mindfulness, are not working, there may be several reasons for this. Those of us who have some sort of history of trauma may have a tendency to disassociate, effectively to 'leave our bodies' because they have not in the past been a safe place to be. We will need to work that much harder at becoming embodied. For grounding to be really effective we will need to heal the roots of that trauma. Linked to this there could also be energetic blockages, which make it difficult for us to earth the energy.

For years I was oblivious to the trauma I was holding in my body, although I became increasingly aware that I could not get grounded through my legs, standing up, like most people seemed to be able to. I found that lying down, feeling the contact with the ground along the whole length of my back was the only way I could even begin to feel grounded. What I came to realize was that there were blockages in the pelvic area that were causing the problem. Most of the sexual wounding that caused these had happened over that series of past lives, which is why I had previously been unaware of the problem at a conscious level. Over numerous years, a combination of the past-life healing and trauma releasing as well as practices such as the Body Scan, have made an enormous difference. Now when I start to open up to the energies, I do not spiral off out of my body, which means I am able to hold and contain a great deal more of such energy safely.

So if getting grounded seems to be particularly problematic for you, it may be worth exploring what else might be going on. Are there other therapies or body work that could help, once the worst of the crisis is over? Again, remember that the height of crisis is not a good time to be doing deep healing work. You need to focus on just getting through it first.

3 – Coping With the Fear

When I was in crisis in Egypt my Mindfulness practice helped me to observe the antics of my mind. I was able to watch how the fear was impacting on my thoughts. Although I did need to take mild tranquillizers for a few days, it was fascinating watching the impact of those too on my mind and thoughts, watching the fear subside.

My particular fear, and it will be different for each of us, focused on sleeping. I knew that our spirit or astral body at times separates from our physical body. This is what makes out-of-body experiences and astral travel possible. I also knew that this separation happens every night when we are asleep. My fear was that if I fell asleep, if I allowed myself to sleep, that I would not be able to come back into my body. What I was correctly sensing was that 'I' was in 'danger', that my sense of self, my hold on my self, was pretty tenuous. It was a reflection of how ungrounded and disconnected from my body I was feeling. It was also to do with the fact that being asleep was the one time that I could not be on my guard and my ego was feeling a very great need to be on guard, its very existence felt threatened.

What I learnt is that fear is very much a creation of the mind. When we understand this at an experiential level, by stepping back from it and catching it at its games, then we can see through it, disarm it. Kimberley, too, discovered this: "My experience wasn't inherently frightening; it was my own fear that almost destroyed me.

Much of how we use Mindfulness to help us work with the fear is based on being in the present moment. If we are feeling terrified, maybe we think we are going to die or go mad, those fears are based on what might happen in the future. If we can stay in the present moment, in this moment right now, we can feel safe, even if it is literally only for this moment that we can feel safe, followed by this next moment and this one.

We stay in the present moment in the way described earlier, by focusing on what we can feel in the body, on the five senses, right now. If you are caught in abject terror then you will need someone to help you, to talk you into the present moment, so to speak.

The other way in which Mindfulness can help us to work with the fear, is that when the mind is in the grip of panic or dread we tend to catastrophize. This type of thinking is characterized by imagining the worst, always in the future, like the example above of thinking we are going to die or go mad. It is highly charged thinking that often bears little relationship to reality, although at the time we find it difficult to see that. By observing our thoughts and gauging the level of 'charge' associated with them, we can begin to spot catastrophic thinking. As we drop down into the body, finding whereabouts we are feeling the charge, maybe a tight knot in the solar plexus or a feeling of nausea in the pit of the stomach, we can breathe into the sensations with kindness and gentleness, soften around them and allow them to dissipate.

This is not a quick fix. We talk about 'working with the fear' precisely because that is what we have to do, over and over again, as the fears and the catastrophizing repeatedly come up. It takes perseverance and patience, time and tenderness towards yourself and your suffering. And there are times when it is appropriate to take tranquillizers to help cope with the terror, as I did in Egypt. There is no need to feel ashamed or a failure just because your fear levels have become unmanageable. Always take prescription drugs under professional medical supervision.

4 – Coping With Unusual or Painful Physical Sensations

We can apply the same principles to physical sensations that we find bizarre, frightening or painful. As we have seen, the way to be mindfully in the present moment is through the body and the senses. You might wonder, 'If I'm in physical pain, won't bringing my attention to the body make that feel worse?' The opposite is true in fact. What we do not want to be doing is denying or resisting the pain in any way. By taking our awareness to it, we are acknowledging it. In the process you can explore the pain or discomfort, watching how it ebbs and flows, how it is always changing, despite our impression that it is constant and relentless. You may even discover, as you start to open to the physical sensations, that they are not as unbearable as you imagined or feared.

So that we do not feel overwhelmed by the pain and feel that there is nothing else in our experience right now other than that discomfort, we then intentionally look for aspects of our experience that feel pleasant, enjoyable, or at least not painful. At any one time, no matter how much pain we are in, there will always be something positive, however subtle, in our experience. We can then broaden out our awareness to be present with all of our experience, all its richness and depth. Working with pain in this way is a Mindfulness approach developed by Vidyamala Burch, the founder of Breathworks.[116]

As well as being useful with physical pain, this use of Mindfulness is particularly helpful with emotional and psychological pain. In spiritual emergency it can be very difficult to ride the rollercoaster of extremes, which can shift from joy and bliss to apocalyptic loss with the speed of an actual rollercoaster dropping suddenly from the heights. When we are in the clutches of painful or negative emotions, to be able to feel the depth and breadth of our experience, to know that there is more to it than the suffering that feels all encompassing, can be a life-line.

5 – Getting Some Space Around the Experience

So Mindfulness is about awareness, especially self-awareness, of thoughts, emotions and physical sensations. In learning to watch the myriad dimensions of our experience as they all flow in and out, we gain a little bit of distance from them, we step back from them. In other words, we create a little bit of space around them, between them and us, and in doing so they lose some of their charge for us. When dealing with the overwhelming intensity of spiritual crisis that little window of space may be enough to make all the difference between feeling that we simply cannot cope and feeling that maybe we can get through this moment. As Mindfulness is about taking each moment at a time, that is all we need to get through; all we need to be able to cope with is this moment.

Brian Keenan is someone who discovered almost by accident this technique of watching our experience, of taking a step back from it in order to gain some space around it. In 1986 he was taken hostage in Beirut along with John McCarthy. Aspects of his experience during more than four years of torture and confinement, much of it solitary, were not dissimilar to spiritual emergency. The sheer psychological, emotional and physical intensity, the rollercoaster of the mind as it toys with madness, the fears and the tears, it is all there in his memoir *An Evil Cradling*. What saved Keenan from losing his mind in his atrocious circumstances was Mindfulness, without his actually naming it as such.

> I decided to become my own self-observer, caring little for what I did or said, letting madness take me where it would as long as I stood out-side it and watched it. I would be the voyeur of myself. This strategy I employed for the rest of my time in captivity. I allowed myself to do and be and say and think and feel all the things that were in me, but at the same time could stand outside observing and attempting to under-stand. I no longer tried to bruise myself by attempting to fight off the day's delirium or tedium. I would let myself go and watch myself, full of laughter, become the thing that my mind was forcing me to be. [117]

Keenan discovered Mindfulness, this cultivating of the witness, and it was to be his savior. Kate too has found this aspect of Mindfulness, of getting some space around what is happening, helpful. We heard the story of her lengthy 'dark night of the soul' ordeal earlier.

> I've been practicing mindfulness for about four years now and find it works on two levels. The immediate benefit is the creation of a 'buffer zone' around difficult thoughts and feelings when they arise, that helps me see them more clearly for what they are. On another level, mindfulness taps me into that sense of oneness I used to experience via mystical states — but this time in a more grounded, less dramatic way.

6 – Slowing the Process Down

Part of what makes spiritual emergency so difficult to cope with is the extreme intensity of the onslaught. Whether it is disturbing visions flashing before our eyes when we are awake, or strange sensations of our whole body vibrating, or not being able to sleep for weeks on end, or all three at once, we need to find ways of making it more manageable, of slowing the process down. Again, Mindfulness can help us here, if we turn our attention to 'doing the ordinary'. This is precisely what the meditation teacher Jack Kornfield refers to in the title of his well-known book *After the Ecstasy, the Laundry*.[118] Every day activities, such as watering the plants, peeling potatoes or walking the dog can help bring us back to ordinary life. With Mindfulness, they can help slow the process down and help us feel less overwhelmed. This needs to go hand-in-hand with easing up on your spiritual practices, ideally stopping altogether temporarily until things settle down. So whether you chant, pray, meditate or practice Chi Gung, focus on letting go of these for now and 'doing the ordinary' as much as possible. I know it is really hard to do this, because our longing to transcend the material, physical realm of everyday life, to unite with Source, God or Allah, is so strong at this time. But rest assured that you will be able to come back to your spiritual practice when it is much safer for you. So where meditating would speed the process up, would encourage the build-up of energy, mindfulness practice and 'doing the ordinary' will slow it down.

These last two aspects, gaining some space around our experience and slowing it down are closely related. In fact all these various ways in which Mindfulness can help us to cope are inter-related.

7 – Surrendering to the Process

One of the biggest and most common mistakes we make during spiritual emergency is that we resist the experience. While this is totally understandable, it is not helpful. The more we want it not to be happening, the more we want it to go away, to just stop, the more, ironically, we will struggle with it. As Brian Keenan discovered, trying to fight the reality of our experience is 'bruising' and he gave it up.

So how do we surrender? How do we stop fighting and resisting? What does Mindfulness have to do with it? When we are ready, then we can surrender. We can stop avoiding our experience, stop running away from it and we can turn to face it. If we take this further, we can even step towards that which is painful or difficult. We can say 'hello' to it, befriend it. We can lean into it, like leaning into the wind, we can soften around it. This is one of the principles of Mindfulness developed in the '5 Step Process' by the Breathworks team and is part of what I teach my Mindfulness students.

In spiritual emergency this might include acknowledging that, no matter how excruciating what we are going through, it is part of a positive process of healing, of clearing out, of growth. Maybe we cannot quite get to the point of welcoming the sleepless nights or the inner turmoil, but maybe we can at least stop resisting them, bear with them. Sometimes it is not until we are so exhausted that we cannot fight or resist any longer, that we are ready to surrender. But we do not have to leave it that long.

The moment of surrender is life changing. It is the point at which we say 'I cannot control this' and we hand ourselves over, we entrust ourselves to something bigger and wiser than us. This is Kate's take on it:

> "The 'surrender' word is a bit of a cliché in spiritual circles but I can't think of a better one. It's like being a puppet on strings, learning to trust an unseen string puller and go with the flow. That's not easy, but in another paradox, the more you relinquish control, the more 'in control' you become in a truer sense of the word. It's easier to accept the ever-changing flow of life by living not in the past or future, but right here, right now — accepting, enjoying, enduring and releasing on a moment by moment basis."

We are so used to thinking we are in charge, that the idea of the process maybe having a logic and a wisdom of its own, far greater than ours, seems outrageous. Yet it is only by surrendering that we come to learn to trust, to trust the process, to trust that we will be okay.

The whole experience of spiritual emergence and emergency is about our awakening, feeling the connection with the Divine, God or Shiva and feeling our own divinity, our own Buddha-nature, coming to know that it is always there and will always be there. It is in that moment of surrender that our deepest longing is met, that we finally touch base with the reality that God, as we understand it, and ourselves are one and the same. It is in the moment of surrender that all the ego's protective boundaries fall away and we experience ourselves as Divine Consciousness. And of course, there is not just one moment of surrender; we keep surrendering, over and over.

Only we can surrender. Nobody can do it for us. And for all our talk of support, there comes a point when we are on our own, when we have to be alone, in order to let go into that connection with the Oneness, the Divine as we see it. Surrendering to the higher intelligence of the process of spiritual emergency and surrendering to the Universe are one and the same.

A word of warning: in some traditions the process of a spiritual teacher guiding you through awakening involves the total surrender of student to guru. If you have any doubts whatsoever about the ethical behavior and integrity of your teacher, then such surrender would not be appropriate. Ultimately the only being we are surrendering to is our Self.

Take My Hand

Mindfulness, cultivating the witness, can help us to spot when we are resisting our experience. A helpful rule of thumb is, if you want it to be different in the slightest way from how it is, then you are resisting. The opposite, the key Mindfulness principle, as taught by Breathworks trainers like myself, is to move towards that which is challenging or uncomfortable, as best we can, with kindness and gentleness towards ourselves. If I had to give just one piece of advice to someone in spiritual emergency, it would be 'trust the process, surrender to it'. Brian Keenan discovered that far from going mad, as he surrendered to the madness, it enabled him to rise above it. I will leave you with my poem to reflect or meditate on:

> I surrender,
> I surrender,
> I surrender,
> In this moment
> And in this moment
> And in this.
> Take my hand
> And we will surrender together,
> In this moment
> And in this moment
> And in this.

CHAPTER 9

Phase 1
Coping with the Crisis
— Looking After Mind,
Body and Soul

The most pressing need is to get safely through the worst of the crisis itself. This is an extraordinarily intense time on all levels, so we need to look after ourselves on those many different levels, physically, emotionally, spiritually and more. In this chapter we will look in very practical terms at how to cope.

SUPPORT

If those in the housing market hammer home the importance of the three 'L's, location, location, location, then for spiritual emergency I must insist on the three 'S's, support, support, support. It is virtually impossible to navigate the treacherous waters of spiritual crisis without support. This in itself, of course, can bring up all sorts of issues around our ability to accept or receive help and also how easy or difficult we find it to even ask for help. Do not leave asking for support too late.

The most obvious and immediate support can come from family and friends, and from your spiritual community, if you have one. I use the term spiritual community very loosely, from an actual faith community to your circle of spiritually committed friends. Because of the level of support you are likely to need at this stage, which may well include having someone with you 24 hours a day, the wider a network of family and friends you can call on the better. You will need to balance this with gauging who you feel safe with, given the level of vulnerability you are likely to be experiencing. If family and friends can coordinate, in consultation with yourself, who is going to be with you when, then that would be ideal. Or you could ask different people to do different things, one person could do your shopping and others could take it in turns to cook, for example.

SUPPORT DO'S AND DON'TS

Do
- Actively seek the support that feels right for you
- Co-ordinate family and friends to take it in turns or do different things
- Look for appropriate professional expertise - ask what experience the professional has of working with spiritual crisis
- If there's a spiritual emergence(y) network in your country, find out what they can offer
- Seek appropriate medical advice when necessary

Don't
- Don't leave it too late to ask for help
- Don't worry about being a burden on family and friends. That's their responsibility to look after themselves and let you know what they can and can't manage

Holding

Part of the reason we need support when going through spiritual emergency is that we need 'holding'. We need those around us to help contain the experience. Because of the intensity of what we are going through, we are probably not going to be able to hold it or cope with it ourselves. This is why even knowing that friends are thinking of you, maybe lighting a candle for you, can feel so supportive. It can provide some essential holding.

There are times, however, when we need more holding than family and friends can offer. This is when we need to think about some residential support, somewhere that is set up for 24-hour care, if that is what we need. There are, unfortunately, very few places that can offer this other than mainstream psychiatric units. At the back of this book, under Resources, you will find a list of appropriate places to stay in the UK, identified by the Spiritual Crisis Network. Most are only suitable once you are over the worst of the crisis and are recuperating, but one or two are able to offer 24-hour support, depending on your circumstances and your financial resources.

If you do go to hospital, ask to see the chaplain, as these days they tend to have a broad multi-faith approach and may be a source of support to you whilst there. Maybe be wary about discussing the spiritual aspects of your experience with any of the other staff unless you know they have a spiritual practice or faith of their own and are therefore hopefully less likely to pathologize what you talk about. If you feel the need to communicate what has been happening to you, it is usually safer to get hold of a pen and paper and write about it keeping that to yourself for now, rather than indiscriminately talking about it.

IF YOU ARE IN HOSPITAL

- If at all possible, get out in nature, even if there is just a little garden area.
- Make the most of any physical activity available, if there is a gym or yoga sessions or something similar. That can help you with embodiment and grounding.
- If it is available to you, also participate in any arts and crafts sessions, through the occupational therapy department. Any form of creative expression will be helpful and hopefully enjoyable.
- If there is a bath on the ward, ask someone to bring you in some lavender oil and have a long relaxing soak.
- Many wards now have quiet rooms set aside for religious observance or practice, such as prayer. That room might be something of a refuge, a safe haven for you. Sometimes staff use these rooms for meetings because of the shortage of space. It is your right, however, to be there. If you have your own bedroom this is not such an issue.
- If you need or want to talk to someone ask to see the chaplain. He or she will hopefully be better able to listen to your concerns about spiritual issues than other staff, who unfortunately may pathologize what you say, seeing it as proof of your 'illness'.
- Be discerning about who you talk to about spiritual issues and how much you say. If you feel a desperate need to talk about what is happening, to get some of it out of your system, it is probably safer to ask for pen and pad and get it all out that way, down on paper.

Professional Support

Another very good source of support and holding, again depending on your financial position, is that offered by a whole range of independent professionals. Transpersonal counselors and psychotherapists should have an understanding of the process of spiritual emergence and emergency. Bear in mind, however, that it may be a difficult time to be building a new trusting relationship with a stranger. Your fear levels may be such that it is difficult to feel safe with anyone. Trust your intuition. You will be very open and very sensitive right now and probably more in touch with your gut instincts than usual. Use that as a powerful resource in this whole process of reaching out for support. In the UK, you can find a transpersonal counselor or therapist who is geographically near you through the BACP (British Association for Counsellors and Psychotherapists) and UKCP (United Kingdom Council for Psychotherapy). In the US, the Spiritual Emergence Network lists available therapists and in Canada the Spiritual Emergence Service has a national directory of therapists on their website (see Resources). You will need to check any one person's experience of working with spiritual emergency.

The focus of individual sessions whilst you are going through the worst of the crisis should be on helping you cope with the immediate, helping you to get and stay grounded, helping you to cope with the fear. It is not appropriate at this stage to be conducting a 'standard' session. Some argue that it is not possible to work with a therapist at this stage of the crisis, but that is because they have not made this important distinction.

In 2003 when I was in crisis, I was lucky to already have a strong relationship with a therapist. During the sessions at that time we focused very much on grounding. We worked in the garden, sitting on the grass, on the earth and with the trees, feeling their rootedness in the ground. My therapist also leant me a large, palm-sized stone. It was solid and heavy. Every night, instead of a hot water bottle, I took the stone to bed with me. I went to sleep holding it, feeling its solidity and weight. For several weeks it was literally all I had to hold on to. This was very different work from what we had been doing and from what, in time, we were able to turn our attention to again.

If you can find a therapist who has been through the process of spiritual emergency and emergence themselves, as mine had, even better. Do not be afraid to ask.

Spiritual Emergence(y) Networks

There are several groups that can be an invaluable resource. They are often run by a combination of mental health professionals and those who have been through spiritual emergency, usually on a voluntary basis. The urge to be of service in this way comes directly out of the experience of crisis itself.

It is worth looking at various websites, not just at your country's network, because they tend to have useful resources. The American and Australian SENs, the Canadian Spiritual Emergence Service (SES) and the UK Spiritual Crisis Network (SCN) can all be contacted by email. The UK SCN offers information and support by email and has a few local groups around the country. The intention over time is to build up the network of local groups around the whole country. When people are looking for support they tend to need it in their locality.

To give you an example, when we are supporting someone through my local group of the Spiritual Crisis Network, two of us always work together when we meet with the person. That way we are able to offer more holding and we can also support each other. We also make sure that of the two of us, one is a health professional and the other has personal experience of spiritual emergency. Many of us in the SCN have both personal and professional expertise. The sessions, an hour long, always focus on support; what support the person has and what more they might need. Unfortunately, what we can offer is very limited, only a maximum of three sessions. Afterwards the person is very welcome to join our monthly group meetings if they feel ready. This is, unfortunately, not typical of what is available currently in the UK, but it is a good model to build on, a model which originated in Glastonbury.

FOR CARERS AND SUPPORTERS

Being well-informed
Learning all you can about spiritual emergence and emergency will benefit both you and the person you're supporting enormously. You will then be in a better position to normalize the experience for them. By this I mean reassuring the person that what they are going through others have also been through, that it is a normal, if very painful and challenging, aspect of spiritual growth and healing.

Avoiding burnout
It is very important that you too, as a carer, get support. Many family and friends get burnt out and it is important to be alert to this very real danger in order to do your best to avoid it. Make sure you take responsibility for looking after yourself, for not overstretching yourself. Call on others for help. You do not have to do it all yourself or manage alone.

LOOKING AFTER YOUR BODY

Our bodies can take a severe battering during spiritual emergency, or, at the very least, be sorely neglected, as our souls aim to transcend our corporeal, bodily existence. Our usual patterns of sleeping and eating will most probably go haywire. That is unavoidable to some extent. The aim, however, is always to work towards re-establishing regular rhythms.

Eating and Sleeping
Whilst your usual meal patterns may have gone completely awry, eating is the one thing that you cannot last long without. Hopefully you will not find yourself in such an extreme situation that you need to be physically fed, like Ramana Maharshi. You may

well, however, need someone to cook for you. If all you manage to do in a day is prepare yourself one decent meal, you are doing well.

It is very important to listen to your body, to eat when you are hungry and rest when you are tired, always, as I said, with the aim of working towards re-establishing your regular patterns. At the best of times, I find that when I need to eat, I need to eat right now and cannot wait. Keep food in the house that can be prepared very quickly and easily. Things like cans of baked beans can come in handy. I tend to find that in times of crisis I also need to eat a lot more than usual because of not sleeping and because I am burning so much off, so bear that in mind too. Again, we are all different, but I find I need more protein than usual, so something like a cooked breakfast, or a vegetarian version of, is ideal. Basically you need solid, hearty food. Living on salads is not a good idea right now.

Things to avoid include all stimulants, like alcohol, tea and coffee, fizzy drinks, excessive amounts of sugar. I made the mistake once of taking a lot of sugar in hot drinks because I thought I needed more energy to keep going (I had not slept for three weeks). All I succeeded in doing was exacerbating the hyper feelings because of the sugar rush.

Whilst a good night's sleep may be elusive, do not underestimate rest as a real alternative. Listen to your body and rest as much as you need to. Resting during the day, maybe listening to some gentle, soothing music is well worth doing. Even at night-time, it is probably better to listen to something and rest, if you cannot actually sleep, rather than getting up in the middle of the night. Again, there are times when it is highly appropriate to take something to help with sleeping, either herbal or, if you need something stronger, sleeping tablets prescribed by your doctor. These are really a short-term measure and if sleeping is a protracted problem over months, you will need to explore alternatives. Many of my Mindfulness students listen to the Body Scan CD at night-time, either to help them get to sleep or as something relaxing to listen to in the middle of the night if they cannot sleep. The Body Scan talks you through, so that as you listen you take your attention to each part of the body in turn, breathing into each area and being in the here and now (see Resources).

Relaxing

The Body Scan is not really designed for relaxation purposes. It is primarily a body awareness practice, but it has the added benefit of being very relaxing. Right now you need as much soothing and pampering as you can get. Soaking in the bath needs to be ull works!

THE ULTIMATE BATH EXPERIENCE

You will need:

- Essential oils to put in the water — for example, lavender is very relaxing
- Tea-light candles to go around the edge of the bath
- A portable CD player to play your favorite gentle music or to listen to a Body Scan
- Your favorite nibbles to eat
- A bath pillow for that ultimate bath comfort

The more you appeal to all of your senses the more mindful and relaxing it will be. Enjoy!

While we are on the subject of baths, if you have been struggling with the burning heat associated with Kundalini energy rising, then cool or even cold baths, or showers, may bring some relief. Be careful not to chill the kidneys by staying in the bath too long. An Ayurvedic doctor will be able to advise you on a cooling diet.

For relaxing, if soaking in the bath is not your thing, then it is worth reflecting on what is. Some of the suggestions under Creative Expression below might feel relaxing for you or some of the grounding activities we looked at. Massage is a great option, both embodying, grounding *and* relaxing.

Relaxing is really important in spiritual emergency. The more you can relax into it, the more you will be able to surrender to the flow, to the process. Managing to relax when we are in the midst of spiritual emergency can, however, feel like a bit of a tall order. You may find that none of the usual things seem to work for you. Just as important can be finding things that may be comforting to us. When we are gripped by terror, with nothing left to hold on to, finding even the smallest thing that can be of some comfort can make all the difference.

POSSIBLE COMFORTS

- A photograph of someone whose love or support you feel or who holds the qualities you sense you need right now. I photocopied the front cover of the book Freedom from Fear, which shows Aung San Suu Kyi, the Burmese human rights defender. I found her inner strength and courage very inspiring and supportive. If you have a spiritual teacher, their photo might feel right.
- Hugging yourself, with the knowledge that the adult in you, the

part of you that can get through this, can hold and comfort your inner child, the part of you that feels helpless and distressed.

- Laughter. Having a cup of tea with a friend who always makes you laugh, watching your favorite sitcom or comedy films.
- Little treats, like bars of chocolate or whatever your favorite snack is. Generally, eating your favorite food is a good idea.
- Listening to bird song, getting out in the sunshine, connecting with nature in small but significant ways.
- Spending time stroking a cat, or dog.

EXERCISING

Exercise has the double benefit of potentially being grounding, if done mindfully, and of helping us to sleep by tiring us physically. You may find you need a lot more exercise than usual, however, in order to get that physically tiring effect. Jack Kornfield tells the story of someone who started to go into spiritual emergency on retreat, after prolonged, intense meditation. As he was an athlete, part of what they did to help him come back down to ground, was they sent him on ten-mile runs every day. Within a few days he was back to normal.[119] On the other hand, it may be that what you personally need is much more gentle exercise, some yoga stretching or walking in nature.

DO-IN

This is a lovely, gentle Chi Gung practice that helps with grounding and stimulates circulation and the meridians.

- Stand with your feet shoulder width apart.
- Be aware of how your body is feeling and relax into it as much as you can.
- Start by lightly tapping the head, just with the fingertips. Be careful not to do this too vigorously.
- Then, with loose fists, gently pat the neck. Keep your wrists loose too.
- Work your way over your shoulders and down the outside of the arms and back up the inside of them.
- From there, gently pat the torso, especially down the sides, and then the buttocks.
- Finally, pat down the outside of your legs, then back up the inside.

Remember to keep your wrists loose and floppy all the time. Enjoy!

Body Work

Other body work is also very helpful, both grounding and embodying, like massage, and therapies that re-balance energies, such as Shiatsu. Other methods include Rosen technique, cranio-sacral therapy, focusing and body psychotherapy. Some of these, like Shiatsu, tend to be very gentle, but can also be very profound, so trust your intuition about whether something is going to be helpful right now in slowing the process down. The alternative is that it might catapult you into further challenges by bringing more tricky material to the surface to process, and you do not need that right now.

Medication

Medication can be quite a controversial subject and it is always difficult and painful when you or someone you love or care for is forced to take psychiatric drugs against their will. Ideally you would be in control of what and how much you take, in consultation with your doctor or psychiatrist. Whilst you might be happy to take mild tranquillizers or sleeping tablets you might draw the line at anti-psychotic medication. If you are anything like Kate, who for years struggled with crippling depression as she went through the purifying cycles of the dark night of the soul, no matter what type or brand of anti-depressants the medics prescribed none of them worked. That is not to say that for some people drugs are not very helpful. Make sure, however, that you are fully informed about the potential side-effects, from independent sources other than the drug companies or your psychiatrist. Emma found the side-effects extremely distressing, what she dubbed 'the Halperidol shuffle'.

A health professional interviewed on Kaia Nightingale's documentary film *Spiritual Emergency* found 'homeopathic' doses of medication to be the most appropriate. This acknowledges that people in this state are highly sensitive and will respond to much lower doses than usual. It also acknowledges that whilst a person in crisis might need help to dampen the experience down enough to be able to cope with it, what is not desirable is suppressing, blocking or artificially stopping this natural healing process. Here we get into the difficult territory again of trying to distinguish between those who are in spiritual emergency and those who are not, which as we saw before is not that easy or that helpful. John Weir Perry's Diabasis research, where patients were given little or no medication, had very positive outcomes. Basically, more research is needed. One problem is that the drug companies wield an enormous amount of power and the research budgets to go with it. It is not in their interests for us to heal ourselves without recourse to medication.

CREATIVE EXPRESSION

The more you are able to find some creative outlet to express what you are going through, the more easily you will move through it and the more quickly you will integrate the experience. Apart from that, amidst all the unpleasant aspects of what you are going through, creative expression can be enjoyable and fun, a temporary release from it all.

We do not have to be a Blake or a Beethoven, a Van Gogh or a Von Bingen, to be able to find a creative outlet for the themes, mythological or otherwise, that absorb us during spiritual crisis. Whether it is drawing or painting, writing poetry or prose, making music, singing, or drumming, craftwork, knitting or needlework, the list is endless of the arts and crafts that can help. They can help to channel and ground the all-consuming energies. As well as acting as a welcome release, they can help us to process and to begin to make sense of what we are going through or have been through. With drawing or painting, in my experience the bigger the paper the better, the more room for freedom of expression; expansive moods call for expansive paper. Clay modeling is another good one. Even your children's Lego will do.

I found writing very helpful indeed, especially in the middle of the night, when I could not sleep. I kept pen and paper by the side of the bed. Writing might be for the sheer hell of it, or it might be journaling the process you are going through or recording and reflecting on the dreams you have. Working with your dreams can be particularly fruitful at this time, as Jung would no doubt tell us. It may be that at this stage all you do is keep a note of them, recording as much detail as you can. Later, as part of the process of making sense of it all and integrating what you have been through, you could maybe find a dream group to join or work one-to-one with a therapist who has the right expertise.

TRY THIS!

Stream of Consciousness Writing
- Get your journal, or pen and paper, and a clock or watch.
- For ten minutes write non-stop, without lifting your pen off the paper, whatever comes to you.
- Do not leave anything out, do not censor it in any way.
- Include anything and everything that comes to mind, even 'this is a silly waste of time'.
- Repeat this daily or as often as you want.
- You will be amazed at what you put down on paper, at what comes through!

Dance and other free form movement, as well as being fun and pleasurable creative self-expression, can also be grounding because they involve the whole body and connecting with the ground. In Five Rhythms® dance sessions it is not unusual to see people slipping and sliding around the floor, draping themselves over the ground. At one of the Revisioning Mental Health conferences I organized, the session everyone was raving about was the Five Rhythms® dance on the Saturday evening!

Again, a small word of caution: be a little selective in what creative activity you choose as some forms of expression are more overtly spiritual than others. Your guiding principle at this time needs to be 'will it open me up even more or will it help me to ground and close down?' Something like chanting, for example, may have the adverse effect of opening you up further. Having said that, I do know someone who found the deep, deep tones of Tibetan chanting to be very grounding. We are all different, so just take care.

A HELPFUL MINDSET

So far we have looked at various different aspects of looking after mind, body and soul during the worst of the crisis. Another important dimension is having a frame of mind that will support you.

A Spiritual Frame of Reference

We need some sort of a context, some sort of a frame of reference, with which to understand what we are going through, a lens to look through. The psychiatrist I saw during my first spiritual emergency, at the age of 20, told me that it was best not to try to understand. He was, I believe, misguided. It was only two decades later, when I had a much more positive experience with which to compare that first one, that I realized not having a framework with which to make sense of what I had initially been through was a large part of the problem. I explored this in a talk I gave at a conference, as I mentioned earlier. My conclusion is that having a spiritual frame of reference has a very positive impact on how a person fares in spiritual emergency. Here is my experience.

When I ended up in hospital, at the age of 20, I did not have a spiritual frame of reference. In fact, at that age, I do not think I had even heard the word 'spiritual'. I had been brought up, like an awful lot of us, in a very secular, even atheist, way. My family was not interested in spirituality at all. So I had no spiritual framework with which to understand what I was going through. The difference was enormous 20 years later when at one point I ended up in a wheelchair for several days, because what I was going through was so extreme. In terms of having a spiritual understanding of the world, by now I had developed a spiritual faith and practice, mainly through my daily meditation practice. I had been on retreat a number of times and been on various courses and workshops. I had been on a whole journey of spiritual exploration, including reading a lot. And by then I had come across Grof's work, which had been an enormous relief to me when I had discovered it. So, by then I had a spiritual lens through which to see and interpret this extraordinary experience. Thanks to that I fared much better.

Validating the Experience

In terms of a helpful mindset, the other attitude that will make an enormous difference is whether your experience is validated or pathologized by those around you. As we have seen, if our experience is pathologized we are maybe told one of several things; that we are ill, we are maybe given a label, maybe given medication, maybe even told that we will need to take medication for the rest of our lives. Essentially the message is that there is something wrong with us. It is a negative message. The opposite of that is when our experience is validated by those around us. The message is totally different, that what we are going through is okay, that many other people have also been there too. Remember that is how Assagioli was recommending therapists reassure clients in spiritual emergency.

Many people in hospital see themselves as being ill, because that is what they are told they are, rather than seeing what they are going through as part of a process of healing and growth. The psyche has such a strong inherent drive towards health and wholeness. If people can see their experience within this broader context, such a mindset helps them enormously.

Let us take a look at someone who gives people just such a supportive, validating message when they start to experience what might be deemed psychotic symptoms. This is the Venerable Ajahn Sumedho, the abbot of Amaravati, the Buddhist monastery. He spoke at one of the Revisioning Mental Health conferences on spiritual emergency I organized several years ago. He told us about what happens sometimes when people are on retreat. Going on retreat is very much a process of spiritual exploration. It can be a very deep, powerful experience for people and they can experience extreme states, including spiritual crisis and what might be described as psychotic symptoms. Ven. Ajahn Sumedho basically reassures people that it is okay, that it is part of a natural process, that it will pass and that others have experienced something similar. The very fact of validating the experience rather than pathologizing it can be enough to make the difference between the person moving through it relatively quickly or the person getting very fearful and stuck in it. This is the same principle that John Weir Perry applied at Diabasis, the alternative crisis home.

For myself, during my first crisis when I was 20, my experience certainly was not validated. Everyone was pathologizing what I was going through; family, friends, mental health professionals. There was no one around me who had a sense that what I was going through might actually be a very important movement towards healing. 20 years later the picture was very different; my experience was validated by those around me. My therapist had personally been through spiritual emergency, so was not fazed by it. He knew that with the right support I could come through and really benefit from it and of course feeling his confidence gave me confidence. The main friend who supported me was also familiar with Grof's work and the literature on spiritual emergency. So I had people around me who understood that however powerful what I was going through, it was also very important. The message I was getting was that this was a breakthrough rather than a breakdown.

Having my experience validated made an enormous difference to how quickly I moved through it. This time I was back at work within two months, rather than losing a year of my life, as I had done previously. It also meant that I could focus on the very positive outcomes rather than the trauma of it all; like a mother giving birth, who is able to quickly let go of the pain of labor, given the miracle of having a new born child in her arms.

FOR CARERS: FAMILY, FRIENDS OR PROFESSIONALS

- What fears do you have?
- Are you genuinely able to validate the person's experience?
- Can you trust the process?
- Do you need to seek support in order to gain the confidence you need?

Understanding What Is Happening

Having the facts and information we need to understand what we are going through also helps enormously. Once I found out that sensations of vibration in the body are quite common, I was a lot less perturbed by them. Likewise I felt very reassured to find out that a lot of people experience electrical equipment going haywire; the hi-fi system playing up for no apparent reason, electric lights going off and back on. I reached the point where it no longer surprised me at all if the bank card machine in shops would not work for me. Denise Linn puts such events down to the fact that our personal bio-electrical field is changing, impacting on the electrical gadgets around us.[120]

The more you can find out about the particulars of what is happening to you the better. This might involve quite a lot of research, which you do not necessarily have the energy or headspace for during the worst of the crisis. If it feels appropriate, maybe ask a carer to help. The more they also learn about what is going on, the better informed both of you will be. At the end of this book you will find a comprehensive list of resources and research tools.

Sometimes we are guided to exactly the right person or book or website that we need. In all the reading and research I had done I had never come across any reference to the sort of seizures that leave you unable to move your body. Just when I needed that information I found myself reading St. Teresa of Ávila's autobiography. Trust the synchronicity; trust the process; and trust your intuition. You are the expert on yourself and what you are experiencing. Do not let anybody tell you otherwise.

A Final Word on Coping with the Crisis

Many of us want to be able to pursue our spiritual journeys safely, without the risk of further crises. Having touched upon another reality, having tasted the freedom of emptiness, having heard the inner silence, it is understandable that we are going to want to deepen into that more. If anything, our longing will be even stronger, our seeking more determined, our hearts more committed.

If we have had more than one challenging period, we are understandably going to feel wary. I feel the answer to this dilemma is Mindfulness. For me, it is a way of practicing grounding and embodiment on a daily basis, whilst at the same time opening more and more to the present moment, awakening more and more each day to what is. Mindfulness is a complete path to enlightenment, to awakening, in its own right. It is only through Mindfulness that I can possibly hope to be earthed enough, grounded in my body enough, to hold the increasing amount of spiritual energy the planet is opening to.

CHAPTER 10

Phase 2
Making Sense of it all —
The Hero's Journey

Heroes are symbols of the soul in transformation

CHRISTOPHER VOGLER

WHAT IS THE HERO'S JOURNEY?

In this chapter we are going to look at using the Hero's Journey to help us integrate our experience of spiritual emergency. The Hero's Journey is a powerful model or map for exploring our life story. It stems from the work of the world-renowned mythologist Joseph Campbell. Through the medium of age-old stories and legends, The Hero's Journey draws on archetypal symbolism and imagery to chart the struggles and victories of what it means to be human. It is an enormously versatile model, which also works phenomenally well as a tool for exploring a crisis of spiritual awakening.

A Universal Model

The beauty of the Hero's Journey is its universality. It works across time and across cultures, tapping into archetypal truths. This universal relevance comes from the fact that it is a metaphor for each individual's inner journey of transformation, healing and remembering. Native cultures have long understood this, using the principles of the Hero's Journey in their rites of passage, rituals and celebrations. Malidoma Somé's compelling book *Of Water and the Spirit* depicts this brilliantly. He tells how he and sixty adolescents went off into the jungle for a month-long initiation, led and overseen by the community elders. What he describes is the ultimate Hero's Journey.

Christopher Vogler's 12 Stages

Taking his inspiration from Joseph Campbell's work, Christopher Vogler, the screen-play expert, outlined what he saw as the principal 12 stages of the Hero's Journey. When I started working with the Journey, I was fascinated by how perfectly Vogler's 12 stages encapsulated what I had been through. I then realized, of course, that it was not just me and my story, but that the Hero's Journey can be found within everybody's story. It is truly universal. So much so, that it will be relevant for you, whatever you have experienced. Whether you have had a particularly extreme crisis to cope with, or whether you are simply journeying through life, with all the usual ups and downs, the Hero's Journey will help you to gain a deeper understanding of your path.

Making Sense of Spiritual Emergency

In spiritual emergency the whole period of integration can take many years. Finding the language to articulate the process we have been through can help us to make sense of it and to ground the experience in reality.

The Hero's Journey expresses the different stages we go through during a crisis of spiritual awakening. Once we have recovered from the most challenging part of the crisis, we may well be left with a need to make sense of it. Or we may feel stuck somehow, unable to go back to our previous way of life, yet unable to move forward or unsure how to do so. Or there may be a particularly painful aspect of what we have been through that we need help with.

Working with the Hero's Journey can help us to process the especially difficult parts of our experience. It can help us to get unstuck, so that we can move forward and step fully into our power. It can help us to make sense of all that we have been through, so that we can take the learning from it, take the beautiful gifts, and let go of what does not serve us. Even if you went through crisis many years ago, working with the Hero's Journey may well be the final piece in the jig-saw to really help you understand what it was all about.

For me it was a way of getting closure on that chapter of my life, at the same time taking the enormous benefits, the Elixir, forwards with me. Before we look in detail at how to work with the Hero's Journey, let me say a little about how I used it and why I believe it to be such an extraordinarily powerful tool for healing.

A PERSONAL ACCOUNT OF
USING THE HERO'S JOURNEY

Why did I decide to work with the Hero's Journey? I was feeling stuck. I was on retreat when I suddenly realized that I had not let go of what I had been through, that I had not fully processed it. I was also feeling burnt out and my physical health was not good. All indications, I felt, that somehow I was off course. I had to do something, because otherwise, before long, I would grind to a complete halt. I decided to take three months out, to give myself the time and space to work on processing it all. I felt the best way for me to do that was to write about it. So I resolved to record my story.

I scheduled the sabbatical for January, February, and March and not long before Christmas I came across an article about Christopher Vogler's work and his 12-stage version of the Hero's Journey. It sounded just perfect for my purposes and I knew that the Hero's Journey was recommended to help with spiritual emergency by the likes of Grof. By some lovely synchronicity, Vogler's book, *The Writer's Journey*, found its way to me. It is a distilled version of the Hero's Journey. Instantly I could see that it would be a great map for exploring the territory I had covered. Soon after a copy of Joseph Campbell's classic *The Hero with a Thousand Faces* also found its way to me.

I spent my days reading and writing, getting my story down. I had unwittingly chosen the best months for that kind of inner work, the winter months. It was now nearly three years since the major spiritual emergency I had had in the spring of 2006. The following year I had organized the third Revisioning Mental Health conference, entitled *Spiritual Emergency: the Return Journey*. Unconsciously, I had already started using the language of the Hero's Journey. I was looking for answers about how to live my life after going through such profound experiences.

Now, as I worked with the Hero's Journey, I saw that it was indeed the Return part of the Journey that I was struggling with. In their own way, the final stages of the Journey are just as demanding as the earlier stages. The challenge is to be able to live in both worlds, the everyday and the transcendent, and to know that they are in fact one and the same.

It was the very last stage of the Hero's Journey, Return with the Elixir, which struck a chord with me. As I reached this stage, I understood that this was where I was feeling stuck. One of the dangers of this final stage is that the learning and the blessings, the Elixir, can get lost, forgotten in the busyness of everyday life. I felt that had been the case with me. Part of the problem was that I was unable to grasp the qualities or the learning I had brought back from my Journey. I needed to spell out for myself exactly what Elixir I had returned with.

What had I learnt? What Elixir had I brought back? I set about recording the most important elements. I had learnt of my new-found inner strength, which I did not fully appreciate until my partner was struck down with encephalitis, a life-threatening inflammation of the brain. I had learnt that the Universe will look after me, in ways that I could not possibly imagine or anticipate. The deep trust this gives rise to is, for me, the path to total surrender.

Another aspect of the Elixir for me was becoming aware of and feeling my vast inner light. We all have this, but few of us experience it so directly and know it to be the fundamental truth of the core of our being. I was also able to strengthen my commitment to be of service, thanks to visions I had at the height of the crisis, during the stage of the Journey known as the Supreme Ordeal.

In assessing my personal Elixir, I saw that I had forgotten, or even been unaware of, the key elements. I saw how difficult it was to have a sense of them in my everyday life, how difficult it was to integrate and ground the qualities and the lessons and take them out into the world. I was not living my life according to all I had learnt. I was not finding it easy to integrate my learning into the everyday.

So working with this final stage of the Hero's Journey was a two-step process for me. First, I needed to get really clear about the make-up of my personal Elixir. Then, I needed to find a way to consciously bring that into my life. This is where the idea of the ceremony was born, to help me do that.

A friend, who helped me put the ceremony together and who led it, is an Inter-Faith Minister. I chose the May full moon because it is when Buddhists around the world celebrate the Buddha gaining enlightenment. It is a powerful full moon, not least because nature's energy is bursting forth with the new growth of spring. I chose a spot in the woods near where I live where we often have campfires. The central prop was a chalice. One by one my friends, members of the local Spiritual Crisis Network group, symbolically poured the different qualities of the Elixir into the chalice. First inner strength was added, then commitment to my vision, then inner light and finally trust in the Universe. By the time the chalice came back round to me it was full to the brim.

I sipped the Elixir and shared it with the others. Quite a formidable Elixir! And it is contained in a chalice that never runs dry. It is up to me to remember to drink from it on a daily basis, to allow it to inform my choices, to be uplifted and inspired by it, to not forget all that I have learnt and brought back from my Journey.

The intention with this beautiful ceremony in the woods was to help me integrate those qualities and bring them into my everyday life. It also marked a stage in my Hero's Journey, the closing of one chapter and the beginning of another, of sharing my learning, the hard-won Elixir, more widely in the world.

THE HERO'S JOURNEY

Remember that the Hero refers to every one of us. Whether we have been through a major crisis or not, whether we are female or male. At first it may not come easily or naturally to think of ourselves as a Hero. There can be no doubt, however, that we are at the center of our own story and, from that point of view, we are the Hero of the story. In the same way that we all carry a spark of the Divine within us, or if you prefer, we all have inherent Buddhahood, so does each of us carry the archetype of the Hero.

In working with the Hero's Journey, I have found the following 12 stages enormously helpful and offer them to you here, with deep gratitude to Campbell and Vogler for their work.

JOURNALING THE HERO'S JOURNEY

As we go through the stages you will find, in easy-to-use boxes like this, suggestions for exploring the Hero's Journey through journal writing. Whilst these are primarily aimed at those of you who have been through and are recovering from spiritual crisis, anyone can make use of the model to explore their life journey. We all have times of major transition in our lives. Working with the Hero's Journey can help you to process these, get closure where it is needed and take steps to move the Journey forward.

If you have been through spiritual emergency, I have designed this journaling process using the Hero's Journey especially for you. I have personally found it enormously helpful and highly recommend it. Do make sure you get some support, if you feel you need it. Writing about what happened, however long ago it was, can take us right back there. Whilst that is helpful to process the experience, we may need help with that, either from a trusted friend, counselor, therapist or someone else.

THE 12 STAGES

The **Ordinary World** is the starting point, the everyday reality of the Hero's life before they are plunged into crisis. Through destiny's **Call to Adventure**, the Hero is summonsed to embark on a perilous journey, a journey of danger, a journey of healing and growth. The Hero may falter. Their fear may result in a **Refusal of the Call**. Through **Meeting the Mentor**, however, the Hero may gain the confidence to go forward,

Crossing the First Threshold, the first challenge or crisis point. Along the way there will be **Tests, Allies,** and **Enemies.**

With the **Approach to the Innermost Cave** the sense of danger starts to grow in the lead up to **The Supreme Ordeal.** If the Hero succeeds in overcoming **The Supreme Ordeal** they will benefit from a considerable **Reward.** The Hero then starts on **The Road Back**, a difficult journey in itself. They meet with one final challenge, **Resurrection**, to test whether they have really learnt the lessons. The final question, when they **Return with the Elixir,** is: will they share it with others, for the good of the wider community? Let us take a look now in more detail at the 12 stages.

ORDINARY WORLD

The Hero's Journey begins in the Ordinary World. This is where we first meet the Hero. The Ordinary World is the everyday world that the Hero inhabits before the Journey begins in earnest. It is the starting point of the story. It is the world the Hero leaves behind, familiar and safe, as they are catapulted into the territory of spiritual emergency.

For Annabel the Ordinary World was living in London, working as a nurse. For Kimberley it was studying at college. For Emma it was living in Bristol, making documentaries for the BBC. Whatever shape the Ordinary World takes, it will have familiar elements, daily routines, aspects we all recognize as part of normal life. In spiritual crisis this all gets stripped away, so that only the remnants of the Ordinary World remain, barely recognizable as such. This is how Jennifer's Ordinary World looked.

> I had recently got my Ph.D., had an enviable job, had become involved in a relationship, lived near my family (and not too near), bought a house in my ideal community, begun a dance group, and found a faith community in which I felt at home for the first time (Quakers). After much hard work, my life seemed to be coming together.
>
> *JENNIFER*

Then comes the Call to Adventure. The Hero is challenged to leave the Ordinary World behind. Its safety and familiarity are soon replaced by an extraordinary world that we could not have imagined even in our wildest dreams. It can be quite shocking how suddenly the known descends into chaos, how quickly the trappings of security get stripped away.

FOR YOUR JOURNAL

What did your Ordinary World look like? Were you working? Studying? Were you single or in a relationship? Where were you living? As you describe it, look for clues suggesting that change was on the way. Can you spot any themes? Can you see now any early suggestions of what was to come? As you look back at that Ordinary World as it was, how do you feel now? Are you still in the process of putting a new life together for yourself? A different, better life? This can take quite some time. Be patient and gentle with yourself.

CALL TO ADVENTURE

This is the call to embark on the Journey. It can come in many different shapes and forms. It might come as the loss of a loved one, of one's health, even of one's mind. As Joseph Campbell says, destiny summonses the Hero. Jennifer had a powerful experience of this, having felt that her life was really starting to come together.

"It all has to go," the voice said in the autumn of 1992. I don't think so! I went kicking and screaming all the way. And yet it all went, much in about six months and the rest within a couple of years. Everything in my life changed. Family members and friends died, relationships ended and changed; it all changed. I ended up fairly traumatized, knowing that something was happening to me that the religion of my youth would have called evil and my profession would call psychotic. Neither of these frameworks suited me.

JENNIFER

Each one of the different triggers for spiritual emergency we looked at in Part I represents the Call to Adventure. Whether that is a painful divorce or giving birth, an accident or surgery, an intensive workshop or intense spiritual practices, the Call to Adventure can take myriad forms.

For Annabel the call to engage on the Hero's Journey quite literally involved a journey as she travelled to India. She felt drawn to that country and, in planning her trip, responded to the Call. For Kimberley, the Call to Adventure came with the death of her mother. Very often the Call itself can be very painful and difficult to cope with.

FOR YOUR JOURNAL
What shape did the Call to Adventure take for you? What life event, loss or situation called you to go forth on the Hero's Journey? Maybe several things combined to become your personal Call. As you explore this be kind and caring towards yourself. Loss of any kind is painful, all the more so where a lot has been stripped away quite quickly. How long ago was that? Remember that it takes years, rather than months, to process and recover from spiritual emergency. How do you feel about that Call to Adventure, that event or loss, now?

REFUSAL OF THE CALL

Because the Call to Adventure so often involves painful loss of one kind or another, the Hero does not usually cooperate willingly. Jennifer's reaction, of going kicking and screaming all the way, is very typical and very human. Yet the more the Hero resists the more painful the process is. For myself at the age of 36, not having yet had children, I was faced with a hysterectomy because of large fibroid tumors. I tried every possible alternative to try to save my womb. The refrain that went round and round in my head, over and over again, was 'I don't want this to be happening to me'. Because the grief was so painful, I found it very difficult to accept, to let go. In hindsight, it was a very healing time. It propelled me to explore and resolve the sexual wounding that had given rise to the fibroids.

Refusal of the Call will not generally get the Hero very far. When destiny summonses they do not have much choice. Any temporary reprieve will be just that, temporary. The irony is that in their ego-driven state the Hero believes they have a choice. The refrain 'I don't want this to be happening to me', whilst understandable, somehow misses the point, that it is happening. Here the Hero needs all the compassion towards themselves that they can muster. It is happening. That is how it is. The suffering requires us to be compassionate towards ourselves, kind and gentle. We can either accept it — which most certainly does not mean behaving passively — and do our best to go with the flow, surrendering to the process, or we can exacerbate our suffering by trying to fight the reality of what is happening.

Another reason the Hero may Refuse the Call is out of fear. They may sense that danger lies ahead. On a soul level they probably already know what is to come and are not very enamored with the prospect. Or they may sense that once embarked upon the Journey there can be no turning back. So the Hero may need help in overcoming their fear or reluctance. Or the Call may need to be repeated more loudly or clearly, maybe with further loss, before the Hero is ready to heed it.

We have seen how hospitalization and the over-use of medication can interrupt the process, stopping it in its tracks. It may not be so much an individual Refusal of the Call as one dictated by the mental health services. For some people the healing process of the Hero's Journey is temporarily or permanently thwarted by the psychiatric system. The Call to Adventure initially came for me in the shape of my parents splitting up during my first year at university. That initial Call was interrupted by the trauma of hospitalization. The Call to Adventure was then repeated over ten years later when my own marriage broke down. This time I was able to embark on the Journey fully.

Sometimes, the Refusal of the Call is so brief, so temporary that we barely notice it. It may be a fleeting thought, an emotion that we glimpse. It can feel almost as if we are catapulted straight from the Call to Adventure into spiritual emergency. Given half a chance we would have resisted, we would have loved to Refuse the Call, but it all happened too quickly.

FOR YOUR JOURNAL

What has your experience of the Refusal of the Call been? When the Call to Adventure came did you resist? A lot, a little? What form did that resistance take? Was there any fear? Write about how you experienced that fear at this stage of the Journey. What have you learnt?

MEETING THE MENTOR

Often the Hero meets the Mentor before they have even fully embarked upon the Journey. The Mentor may take the form of a therapist, a healer, a health practitioner, or a spiritual teacher, to name just a few of the possibilities. They may even take the guise of a kindly, wise boss in the work place. And the Hero may have more than one Mentor who helps them prepare for the Journey.

Meeting the Mentor will aid them in moving through the fear and resistance they may be experiencing. With their support and guidance the Hero gains the confidence to be able to engage fully with the Journey. They may receive a gift from the Mentor, a symbol or reminder of their support.

My key Mentor was the Italian spiritual teacher. He is associated with the emblem of the white rose and is holding one in the photo I have of him. At a critical juncture of my 2006 crisis a beautiful bouquet of large white roses found its way to me. At the time it was a palpable reminder of his support, support that I desperately felt the need for. It helped give me the strength to hang on in there.

The Mentor, then, represents the protective forces of destiny. Meeting the Mentor at this stage is no accident. The Hero needs all the help they can get to prepare for what lies ahead and to gain the knowledge and courage needed to move forward. Ultimately, however, we, the Hero, have to go it alone.

Can you identify your Mentor? Is it a man or a woman? What does that mean to you? Have you had one Mentor or several? In what ways did you feel their support? How did Meeting the Mentor help you to over-come fear? Looking back now, do you have a sense that the Mentor represented the protective forces of destiny? How?

CROSSING THE FIRST THRESHOLD

The First Threshold is the first real challenge, the initial crisis point of the Journey. It is at this stage that we engage fully with the process. Often it is only with hindsight that we can identify the First Threshold. What at the time I thought was a full blown crisis, bearing all the hallmarks of a classic spiritual emergency, turned out to be just the First Threshold.

Crossing the First Threshold helps us to prepare for what is to come later. In coping with it we learn a great deal that stands us in good stead for further along the Journey. As we Cross the First Threshold and rise to the challenge our ability to 'bear with' extreme circumstances expands. We become a bigger container, able to hold more of our experience.

If aspects of a person's Ordinary World did not start to fall away with the Call to Adventure, then they are likely to at this stage. There may be a redundancy, a break-down in a marriage or long term relationship, a health crisis, or the loss of one's faith, to name just a few of the infinite possibilities.

The Hero's Journey is as much an inner Journey as an outer one, the internal and the external reflecting each other. So there will also be material coming up from the depths of our psyche, wounding coming up to the surface for air, so to speak, for heal-ing. At this stage we may also have to start dealing with some of the more extreme physical, emotional and energetic effects of the process that we looked at in Part I. We may not have slept for nights or even weeks on end. We may be experiencing a strange vibrating feeling all over our body. Our emotions may be swinging wildly from one extreme to another, and much more.

There may also be what Vogler calls Guardians, who bar the way, making the Crossing of the First Threshold all the more challenging. If this stage came in the form of a hospital admission, you may have experienced mental health staff as well-inten-tioned Guardians impeding your progress. This was Annabel's experience of being forcibly injected with heavy-duty anti-psychotic medication against her will. Courage and determination are needed by the Hero to proceed.

> FOR YOUR JOURNAL
>
> What shape did Crossing the First Threshold take for you? Did an aspect of your Ordinary World fall away? Were there any Guardians blocking your way? What do you feel you learnt, how did you grow with this first challenge? How did Crossing the First Threshold prepare you for what was to come later?

TESTS, ALLIES, ENEMIES

The First Threshold turns out to be the first of many Tests, all of which prepare us for the Supreme Ordeal ahead. In Crossing the First Threshold and coping with it, we need all the help and support we can get. For Annabel, having had Guardians of the First Threshold to contend with, the psychiatrist who honored the spiritual dimension and recognized the value of medication used sparingly was something of an Ally for her. Other Allies had been the Indian family she stayed with for nine months, who supported her with their kindness and generosity, without judgment or fear. Also, her employers at the time of her first hospital admission, who were very understanding and supportive of the process she was going through, were Allies.

So Allies may come in the form of those offering support in one way or another. We are also likely to come across one or more 'enemies'. These are people who we are projecting aspects of our shadow onto. The parts of ourselves that we do not like and have suppressed or the parts of ourselves that we do not consciously acknowledge, we will see in others. So a person we find intensely irritating is, on some level, reminding us of ourselves. Our good qualities can go unacknowledged too, so a person we admire and maybe feel jealous of is embodying qualities we have but do not see or have not yet developed.

A particularly painful and tricky aspect of spiritual emergency is that someone who has until now been a Mentor or an Ally can now seemingly become an Enemy. As the really difficult parts of our psyche come up to the surface we project them onto those around us. Because we are also experiencing a lot of fear it can become very difficult to trust anyone. Even those we have previously trusted can become Enemies. We maybe get to the point where we are no longer sure who is an Ally and who is an Enemy. This is all part of the Tests we are facing now. Can we discern who to rely on? Can we trust our intuition?

FOR YOUR JOURNAL
Following the first crisis point, the First Threshold, what further challenges or tests did you encounter? Can you see ways in which they too prepared you for later? Who were your Allies? How did they support you? Who was your Enemy? Was there more than one? Were you right not to trust them? What aspects of yourself did you project onto them? Write about what you learnt from your Allies and your Enemies.

APPROACH TO THE INNERMOST CAVE

This is the final stage before the terror of the Supreme Ordeal. The ground is being prepared for that. By now the Hero is likely to be very open energetically. They are also likely to not be very grounded as they are blasted by the spiritual energies coming through. They are operating in another realm now. There is much at stake at this stage of the Journey. Amongst other things, the Hero may be in danger of physically hurting or harming themselves in some way, if they are too out of touch with their bodies and their physical needs. This is where their Allies will come into their own, helping to hopefully keep the Hero safe.

People's experience of the Approach to the Innermost Cave varies. Given the degree of 'openness', the Hero may experience considerable synchronicity, maybe precognition, having an intuitive sense of things that are about to happen. For Kate, the Approach to the Innermost Cave 'was accompanied by the most intense synchronicity I have ever had'. The Hero may also experience a sense of oneness with all things or a strong connection with the Divine. This may be a time of considerable spiritual insight and learning for them.

These are aspects of the Reward stage, which for others may come after the Supreme Ordeal. This reminds us that the Hero's Journey is not a linear process. The Hero may move back and forwards through some stages, skip others, whilst some overlap.

The Approach to the Innermost Cave can be more difficult for other people, as the levels of fear they are experiencing begin to mount rapidly. Things can begin to take on a 'life or death' quality. Realizing that the death the Hero is beginning to fear is an ego death rather than their physical death can help enormously. This is a very important point and worth repeating. When the fear of dying is very strong, understanding that it is the ego rather than the physical body that is dying, will help the Hero a great deal.

Either way, the Approach to the Innermost Cave is preparing the way for the Supreme Ordeal, as the Hero becomes increasingly 'open' and focused on the inner realm. Crucially, the Hero is on his or her own now. This is part of the process. They have to go it alone in order to find and experience their connection with the Divine, the Universe, God or Spirit. This goes hand in hand with finding their inner strength and learning to trust. This does not mean being without supporters, or without Allies. But even if those helpers have been through spiritual emergency themselves, at this moment in time, the Hero will journey alone into the deepest recesses of their inner world. This is the beginning of the descent into their personal hell realm of the Supreme Ordeal.

FOR YOUR JOURNAL

Can you identify your Approach to the Innermost Cave? Was it a very frightening time for you or did it have elements of the Reward? Maybe it was both. How did it lead up to the Supreme Ordeal? Who were your Allies? Write about what if felt like to be so alone with the process. Can you see any positive aspects to that, any learning? As you are writing and reflecting make sure you get the support you need right now. Maybe talk things through with a trusted friend, counselor or therapist.

THE SUPREME ORDEAL

This is the most excruciating and terrifying part of the whole Journey. It is the crisis with a capital 'C'. The Hero comes face to face with their deepest fears, whether of losing their mind, of dying or some other dreaded circumstance.

During the Supreme Ordeal, the Hero always faces death on some level and the outcome is by no means certain. Malidoma Somé, in his book, tells how four of the 60 youngsters who went off into the jungle for their initiation into manhood, did not return. We know too, from the tragic story of the American woman's death, how very high the stakes are. We have also seen how very real the fear of death is for many as they go through the Supreme Ordeal.

A buzzing began in my head. It became a deep, resonant drone and then a thundering, rumbling screech. It was like a freight train crashing through my ears. The noise began to move through my body along with heat, convulsions, cramps and pain. I was terrified. I began to fight against what felt like paralysis, I was trying to move, open my eyes, scream for help. No sound or movement came.

The air I was breathing was now so hot it was scorching my nostrils and I could feel it burning down into my lungs. I thought maybe I was being taken over by something. Whatever was happening had tremendous force and direction and I was fighting against it. I was aware of some other consciousness, just as when I always knew when someone else had walked into a room or was watching me from afar. This first experience felt like a fight for my life. The more terrifying thing was I didn't know what I was fighting against.

When it passed it passed with a beautiful, blissful whisper of breath accompanied by extremely bright white light, as though it had been the most angelic experience rather than one so terrifying. As it ended, my body flopped onto the bed and I realized that for the last couple of seconds my body had lifted off the bed, I don't know by how much I just remember physically landing back onto the bed with a bump, my eyes were open. It felt like every muscle in my body had been torn.

KIMBERLEY

The Supreme Ordeal takes a different form for each of us. Ultimately all are transformed and reborn through the experience. The process of having to face our deepest fears and overcome vast opposition, whilst at the same time letting go into the deepest surrender, helps us, the Hero, to connect with the Divine and discover inner qualities we did not know we had. I remember in my darkest hour feeling the energy of Shackleton, the extraordinary Arctic explorer, who not only survived the unsurvivable but also managed to save all his men from what looked like certain death. I found his presence so very comforting. He helped me to draw on reserves on inner strength I did not know were there and I also remembered that at the most treacherous part of his journey, when all seemed lost, he felt the presence of the Divine.

Each person, in their own way, gets to discover that they can live through the unlivable, the unimaginable. This is central to being transformed and reborn by the Supreme Ordeal. There is a sense of a new being born from the Ordeal. This is the rebirth of the 'death and rebirth' motif so familiar in spiritual emergency.

The next four stages of the Hero's Journey represent the Return Journey. For those, however, whose Supreme Ordeal results in psychiatric hospitalization, the Return can be impossible or extremely difficult. They may get sucked into the 'revolving door' syndrome of repeated hospital admissions. The Supreme Ordeal may be the end of the road for them. The daughter of one couple has been in hospital all her adult years. Her latest treatment has been enforced electro-convulsive therapy.

FOR YOUR JOURNAL

As you revisit your personal Supreme Ordeal make sure you take good care of yourself. If the journaling starts to feel a bit overwhelming, put it to one side for the time being. Focus on nurturing yourself and getting support. You can then come back to it, if you want to, when you are feeling more resourced.

What deepest fears did you come up against? Were you afraid you were actually going to die? In what way did you experience the symbolic death and rebirth? How do you think the Supreme Ordeal transformed you? Write about the inner qualities you discovered within yourself and any sense you had of Divine support being available to you.

THE REWARD

The Reward for overcoming the Supreme Ordeal can come in many forms.

> It has led me to a path of such love, wisdom and gifts beyond anything I would have thought possible eight years ago when I wanted to go to sleep and never wake up. It is my personal belief that during 'spiritual emergence' we are given a glimpse of something beyond the veil of illusion.
>
> *KIMBERLEY*

With this kind of crisis of spiritual awakening, the Reward may be an experience of profound grace. It may be deep insight into the nature of reality. Or it may be some aspect of the Mystical Marriage, such as a balancing of the inner male and female. Often this takes the form of new found love, of a new relationship, the classic Reward of many a story.

> A major symptom of being cut off from myself, was my inability to form fulfilling, long term intimate relationships, which was a painful legacy, when of course a longing for deep connection, for being seen for who we are, lives in all of us. I have recently embarked on a new relationship which is opening up a rich, new world of closeness and sharing of an order I have never previously experienced.
>
> *ANNABEL*

Even those who get caught up in the mental health system may have glimpsed enough of the Reward to struggle on with their Journey. Before ending up in hospital they may have briefly experienced infinite love or total inner peace. Something that was enough for them to know the experience was real and want to find out more, want to pursue it.

The Reward is what makes it all worthwhile, all the agony, the fear and the suffering of the Supreme Ordeal. I have yet to come across a single person who regrets what they have been through. Often the Reward works on many different levels and it can take quite some time to fully appreciate all the different aspects of it.

FOR YOUR JOURNAL

What shape has the Reward come in for you? Has the internal transformation resulted in external changes, such as a new relationship or a new vocation? Or has the Reward come more in terms of inner qualities or realizations? Write a list, as long as you can, of all the different aspects of the Reward. You may be amazed at just how much there is to your personal Reward.

THE ROAD BACK

The Road Back to the Ordinary World is not easy. In spiritual crisis this can be a particularly painful and challenging part of the Journey. Many fail to make it back, to complete this crucial stage. Having visited the Special World, with experiences of transcendence, of bliss, of oneness, the Ordinary World can now feel devoid of meaning and no longer seem that appealing. There is also the pain of separation from the Source or the Divine to be borne. So, before the Hero can even begin to come to grips with the Ordinary World again, they must first make a decision, consciously or unconsciously, that they are willing to travel the Road Back.

Having made that decision, working out how to function in the Ordinary World takes some doing. How can we 'fit in', after what we have been through? How can we make sense of the Ordinary World, its materialism, its upside down priorities?

I gradually understood that, regardless of how much I felt I was getting back to normal, I would never be the same again and any attempt to try to be that person was dishonest and dishonoring to myself. From here on I would have to find another way to function in the world. It took me a while to give myself permission to be different.

KIMBERLEY

It is now a question of learning to straddle or bridge the two worlds. It is a question of realizing that they are in fact one and the same. This adjustment can take a long time and some do not manage it, ending up isolated, cut off from family and friends and society at large. Others are unable to earn a living or pay for a roof over their head. The most unfortunate spend the rest of their lives locked into the psychiatric system.

Others, often over a number of years, gradually find their way back into the Ordinary World on their own terms. They usually have to make considerable changes in their lives. They tend to speak of having found a stronger sense of purpose and meaning to life, of living deeper, richer, more fulfilling lives, often incorporating an element of service. This whole issue of *Going Back Out into the World* is explored more fully in the final chapter.

FOR YOUR JOURNAL

What appeal did the Special World have for you over the Ordinary World, if any? Were you reluctant to travel the Road Back? How did that show itself? Did you make a conscious choice to re-integrate back into the Ordinary World? Or did you gradually come to terms with the need to do that? Do you feel you need more help with this stage of the Journey, The Road Back? Identify who you might be able to get the appropriate support from and get in touch with them.

RESURRECTION

Back in the everyday world, one final challenge awaits the Hero, to see whether they have retained what they learnt, whether they are able to apply their new-found knowledge in this world. At this point in the spiritual emergency experience there may be one final challenge or crisis. It may be a reminder of the death the Hero faced during the Supreme Ordeal in some way. If you have assimilated what you have been through, you may deal with this in a new, different or creative way.

A year after my 2006 crisis, my partner was suddenly struck down with encephalitis, similar in some ways to meningitis. One minute he had shingles and the next the virus was attacking his brain. As I waited for the ambulance in the early hours of the morning I did not know any of this. I did not know how close a shave with death this actually was. All I knew was that something was horribly wrong, that maybe he had had a stroke. I could feel myself panicking, the adrenalin rushing round my body. Then I remembered all that I had been through, all that I had coped with against the odds, during my spiritual crisis. "I can handle this" I thought. "Whatever happens, I can handle this."

At this stage of the Journey the stakes are still very high, because not only may the Hero become stuck between the two worlds, unable to move forward, but the learning and gain for the wider community may also be lost. If the Hero does not remember what they have learnt, if they are not able to apply it in the Ordinary World, all that has been achieved can be undermined at this stage.

RETURN WITH THE ELIXIR

Life for the Hero will never quite be the same again. The end of the Journey marks a new beginning. We are reborn. Others see how we have changed. It may take quite some time, however, for this to become apparent. In spiritual emergency the period of integration can take many years. If the Resurrection stage was one final test to see if we have integrated what we have learnt, then this stage is all about taking that learning out into the world. It is about standing in our power, bringing all of ourselves to all we do, with a level of receptivity and surrender that comes from totally trusting the Universe to guide and protect us.

The Elixir, the learning, the blessing, the remembering, is what we, the Hero bring back to share with others. It proves we have been to the Special World, that the Supreme Ordeal can be survived, death can be overcome. It serves as an example for others.

Will the Elixir be forgotten, lost in the busyness and demands of everyday life? Some try to get on with life as before, hoping to put the whole experience behind them. Others, because they struggle to integrate what happened, are not able to ground the learning into everyday life. Sometimes the resulting sense of calling or vocation feels like it brings so much responsibility that fear makes it difficult to take the called-for action in the world.

Or will the Hero share the Elixir with others? This is the final test. This is what makes them a true Hero. And the Elixir may be so powerful that it has far reaching consequences, sending waves out far beyond the Hero.

The strong impulse to be of service that many feel following a crisis of spiritual awakening is a prime example of sharing the Elixir with others.

The strength I've found as a result of having some profound Dark Night of the Soul experiences during my illness have given rise to a deep devotional leading to be of service to others. For some years I've felt drawn to teach meditation and to train as a Spiritual Healer.

EMILY

THE 12 STAGES TOGETHER

We have seen how with the Hero's Journey we come full circle. The starting point of the Ordinary World is also the final destination, but now experienced completely differently. To paraphrase T. S. Eliot's words in *Four Quarters*, [121] we arrive back where we started from after our exploring, and we see the place as if for the first time.

The Hero now takes action in the Ordinary World based on totally different priorities. Looking back, it is much easier in hindsight, to identify the different stages, from the Call to Adventure, to Meeting the Mentor, to the Supreme Ordeal and the various aspects of the Return Journey.

And there may be further cycles of the Hero's Journey. The Call to Adventure may come again in a different guise. It may, or may not, be Refused and there may be a new Mentor to meet along the way. And so the Journey continues.

As I finished work on this section of the book, taking a break from the final editing, I walked past some open windows. Music drifted out. The song? 'Search for the Hero Inside Yourself' by M People!

CHAPTER 11

Phase 3
Going Back Out
Into the World

After Coping with the Crisis and Making Sense of it All, our next challenge is to find our way in the world again. This is the Third Key Phase; Going Back Out Into the World. In Hero Journey's terms this is the Return part of the journey. Our challenge is to work out how to live with one foot in each world, how to straddle the two realms, in time managing to integrate them thoroughly enough to become one. There are no easy answers. Each of us has to work it out for ourselves, hopefully with a little guidance from those who have gone before.

The sorts of issues we will face include managing to earn a living, whilst following our soul's calling, coping with acutely fine-tuned sensitivity and finding a way to pursue our spiritual practice safely without spiraling off into crisis again.

On a deeper level, we may have to battle with a strong sense of dissatisfaction with the material realm, with a desperate longing, yearning, to live permanently in utter union with the Divine. Do we even want to live in this mundane world? Do we even want to re-engage with it? Are we ready to? The 'yes' may come very slowly. Or it may not come at all. Not everybody chooses to or is able to make the Return journey. Some get caught up in the psychiatric system. Others take their own lives, intentionally or accidentally.

So the Return, Going Back Out into the World, is certainly not without its trials. We need as much kindness towards ourselves in this Key Phase 3 of Moving Successfully Through Spiritual Emergency, as we did in the previous two. At the same time as trying to work out how to be in the world, we may still be grappling with processing the intensity of what happened. Key Phase 2 and Key Phase 3 may well overlap or we may move in and out of them. Just as the integration of Key Phase 2 can take many years, if not the rest of our life, so too can this Key Phase 3.

As Kate finds her feet, having been through ever deepening cycles of the dark night of the soul, she says: "This time I'm under no illusion that it's the end of the process. It never ends, but hopefully it does get easier."

This is echoed by Kimberley:

> I know my awakening is a lifelong process. The initial emergence was traumatic for me and changed everything. There isn't an end point to the growth but there is an end point to the chaos and trauma eventually.

Whilst things will feel easier and less traumatic now, re-engaging with everyday life brings its own demands. We find ourselves back in the Ordinary World, but we are not the same person. We have changed and so have our priorities.

FOLLOWING A CALLING

One of the first things many of us rethink is usually our working life. We may have had a period off work or have had to leave our job altogether, because of the protracted nature of our crisis. You may find you become unemployable in the traditional nine-to-five sense, because of your increased, highly developed sensitivity or because you need to look after yourself so carefully. This has certainly been the case for me. You may feel a need to retrain, to make a complete career change, or at least make adjustments to your working life. Often this now feels like following a calling, maybe one that has been nagging at us for some time. I had known for quite a while that I needed to write before I finally got down to it. In Kate's words:

> Life is presenting a new set of challenges that seem designed to get me back into the world. The chaos of the last fifteen years has impacted on my work life and I need to start over. I feel ready to pursue my 'calling' as a scriptwriter. Writing about the stuff of life involves going to the underworld and until recently I was so overwhelmed by my own process I'd have drowned in introspection. Now I feel out of danger, like a miner who returns safely to ground level after each day's work. In the deal she struck with Hades, Persephone alternated between the underworld and the surface, equally at home in both and I relate strongly to this. If I can act as guide for others in dark places through my writing, so much the better.

Some of us, as we search for what helps us to heal, to look after ourselves, create a whole new working life in line with that. Having discovered how helpful and powerful Mindfulness was as a practice, I went on to train and become an accredited Mindfulness Trainer. That work is now part of my freelance mixed portfolio of speaking, writing, and teaching.

Kimberley too, by finding what helped her personally, felt drawn to use the same tools to help others.

> I found spiritual teachers and information that helped me manage and balance my energy (Chi/Ki/Prana) and these approaches really worked for me. I got strong enough to train in these energy practices, at the same time mastering them for my own ongoing healing, development and strengthening.
>
> I hadn't had a clear career path before my first Spiritual Emergence event in February 1998. There was a general theme of art and I'd been working in a local gallery but I wouldn't have described it as my calling. So you could say that my need to heal and feel well again led me to a new career or calling, that of energy work.

Kimberley went on to establish her own healing practice, gradually building it up at a clinic above the bookshop where she worked part-time.

As we change and our priorities change, so too does our attitude towards work. You may find you can no longer settle for employment that does not feel fulfilling. You may also find that you are less caught up with it, more detached from it. This is very common after spiritual emergency. We find ourselves living in the world, but not of it, not sucked in by the drama of it all. This was Al's experience:

> After Spiritual Crisis I found that my career was far less important to me, both in the sense of the material rewards it offered me and of what it meant for my self-esteem. It no longer defined me and I no longer found the meaning in my work that I had previously. I found this immensely challenging and it took several years for me to find new (different) meaning in my paid employment. My reduced energies also led me to change my work/life balance and I moved to a four and a half day working week, eventually taking an early retirement package.

As in Al's case, often what is new in this equation is prioritizing our own well-being, mainly out of sheer necessity, having been through so much. We may have become so sensitive to energies that there are many environments in which we can no longer work, or we may still need to rest a great deal and not be able to manage a full-length working day.

The combination of now needing to prioritize our own needs along with losing interest in the material rewards of employment presents us with a dilemma. This is one of the greatest challenges of this Key Phase 3: how can we manage to earn a living?

Earning enough to pay the bills is beyond the reach of some, maybe even for years on end:

> One thing I was not able to sustain after my crisis in 1998 was enough of an income to pay my own way and with this came great guilt and some shame. I did try to go back to working in various jobs for around three years, but always had to leave due to physical or emotional symptoms overwhelming me.
>
> *KIMBERLEY*

BEING OF SERVICE

Closely related to following a vocation is the new-found urge to be of service. You may have noticed this, especially the urge to help others who have been through spiritual emergency.

> The concept of service is also becoming important. Never one of life's natural carers, I'm now drawn to work that helps people develop. Long term I hope to gain counseling skills and continue my involvement with the Spiritual Crisis Network. There was no help or advice for people in crisis when I needed it and I feel passionate not to mention duty-bound about helping to change this. Encouraging awareness and new attitudes within mental health services is just as important.
>
> *KATE*

Kate's and Kimberley's feelings are very typical of those who have been through it:

> I endeavor to take good care of myself so I can be of use to others who may be experiencing signs and symptoms of awakening right now. There is a sense of joyful responsibility to use what I've been through to help others and that's where I'm at today, creating different ways to do that.
>
> *KIMBERLEY*

LOOKING AFTER OURSELVES

This need to take good care of ourselves comes up again and again. It emerges just as strongly during Key Phase 3 as it does during the earlier Phases.

> One of the things that keeps coming up for me is the importance of taking good care of myself, especially of nourishing my mind and body. And in my relationships I'm getting better at balancing my own needs with the needs of others.
>
> *EMILY*

Often it is connected with balance:

> It is a constant balancing act, ensuring I stay well and resourced and making sure I continue to acknowledge what I need in order to stay healthy, regardless of what others may think of my choices.
>
> *KIMBERLEY*

Here too, Mindfulness plays an essential role, in helping us look after ourselves. The heightened self-awareness that comes with a regular practice means we can spot that much earlier if we are getting out of balance, if we are over-doing it. We can use techniques like the three-minute breathing space, a mini-meditation, to help us stay centered and grounded. We can use 'pacing', which involves taking timed regular breaks, in order to be effective in all that we do, nurturing ourselves well in the process.

NEW BEGINNINGS

There is a real sense, as we come through spiritual emergency, of being reborn. As we have seen, death and re-birth are central themes throughout. It is not surprising then that this new stage of our lives should also see new beginnings, as we let go of the old that no longer serves us. Our priorities will have changed right across the board, some more obviously than others. Emma puts it succinctly: "I changed the way I shop and eat, retrained in a healing profession and found a soul mate."

As well as our working life, our home life may change. We may need to move to a more supportive environment, such as living in the countryside. If we were in a relationship with a significant other, we may realize that we need to end that. If we were single, we may start a new partnership. Like Emma, some people discover, after the huge shift of spiritual crisis, that they find their life partner or soul mate, attracting them easily into their life. This was Kimberley's experience:

> Those first two years were hard: depression and physical illness, com-
> plete shifts in priorities leading to relationship changes, job changes and
> several house moves. It was chaotic. After the chaos, the dark days and
> then several months of retreat and rest I felt re-born. I was in such a
> good space, so healthy and resourced that I was able to find (attract) my
> life partner. I was able to say 'Yes' to something really good and healthy.

ENGAGING WITH A SPIRITUAL PATH

Many of us who go through spiritual emergency were already consciously on a spiritual path before the crisis. The experience tends to deepen this, to strengthen the commitment for us. Having experienced a degree of awakening, having seen our potential
to go all the way, we engage with renewed vigor and hope. For others the crisis seems
to come out of nowhere; we were not particularly searching spiritually before, but now
we feel drawn to do so.

Emma saw that the spiritual dimension is not somehow separate from the rest of
our lives: "I see my life as a spiritual practice. I feel like I have a Higher Purpose and
when I'm most aligned to it doors open and opportunities fly in."

At the same time many of us find we have to engage very carefully, almost warily,
with spiritual practice, because of our susceptibility to the powerful energies. Having
been blasted open once, or more, things can very easily start to speed up again. Some
of us get into difficulties with this, others learn to manage it knowing when it is time
to ease off with their spiritual practices. Your partner, having been through the worst of
the crisis with you, may understandably be reluctant to see you go off on workshops or
trainings that could re-ignite the process.

Emily has learnt to gauge what she can and cannot do, to be flexible, the key again
being balance:

> My spiritual energies can still be raised very easily so I have to strike
> a balance, making sure I have plenty of earthy, non-spiritual, ordinary
> things in my life too. As well as sometimes having to stop meditating
> altogether for a short while, if I feel myself getting ungrounded, I've
> reluctantly had to give up having spiritual healing, except very occasion-
> ally. I try to avoid practices that are designed to raise spiritual energies.
>
> I have Kriyas[122] that I experience as an incredibly intelligent, loving
> and healing energy. I set aside time each day for these. They first oc-

curred following an out of body experience and I found them absolutely overwhelming initially, but in time things worked out. I managed to integrate them by allowing time for them each day and keeping them to the discipline of those times.

Emily has found a way of staying engaged with her spiritual practice, whilst at the same time being careful to keep a healthy balance, so that she can function in the everyday world. She goes on to say:

> I don't avoid spiritual experiences but nor do I seek them. I'm comfortable with the Buddhist view that spiritual experiences are a side effect of spiritual practice. As Jack Kornfield says, they can bring 'gifts of inspiration, new perspectives, insight, healing or extraordinary faith' but they do not in themselves produce wisdom and must be integrated if we are to learn from them.

CHOOSING SUPPORTIVE ENVIRONMENTS

In order to flourish back out in the world we need to choose not only supportive environments but also supportive activities, people, diet and more. Emma mentioned that she changed the way she shopped and ate and Kimberley too says her 'diet continues to get healthier and cleaner'. To some extent we may find ourselves naturally drawn to what is supportive, but we may also need to make conscious decisions to avoid certain things.

> There are many films, reading materials, music, environments that I consciously avoid now because they bring me down and make my body physically tremble or ache. For me this is normal now. My sensitivity simply appears to reveal harshness and unkindness that others are numb to or medicating themselves against.
>
> *KIMBERLEY*

Personally, I find the energy of places such as supermarkets and pubs very difficult, which means that I avoid them as much as possible. If I do have to be in crowded, public places then a walk in nature afterwards clears my energy again. Almost unanimously people speak of needing this healing contact with nature. This is because it is grounding, it recharges our energy and, as Kimberley says, earth is 'the ultimate ionizer'. Many actually move to more rural areas because of this.

Most also talk of enjoying a simple, quiet lifestyle; consumerism has lost much of its appeal. The daily onslaught of the media is weeded out as we keep sensory stimulation to a minimum.

> Since Spiritual Crisis, I've generally found that living simply and a balanced routine with plenty of mental and physical relaxation helps a lot. Although I'm getting stronger, I'm still more sensitive to negative energies than I was previously and find that relationships and the general busyness, noise and clamor of the world can very easily drain my energy.
>
> *EMILY*

Emma talks about the importance of supportive people in helping her to re-engage with the world at large:

> I've been seeing a Shiatsu Practitioner who disclosed that he'd also experienced a Kundalini awakening in his 30s. Making time to be supported and guided by someone who's got a good understanding of spiritual awakening feels crucial to me. I feel safe to explore and talk about my experiences without fear of being pathologized. Building this trust with the world helps me want to take part in it again.

COPING WITH INCREASED SENSITIVITY

To a large extent it is our increased sensitivity, which makes us less able to tolerate certain environments, noise levels and so forth. Probably highly sensitive people to start with, we now find ourselves even more so.

> My system feels very finely tuned, which serves my creative and intuitive work immensely, but can make interacting with the physical world somewhat challenging. ... If the volume on my physical senses and intuitive senses isn't turned up too high then I can get on with work in my office. Sometimes I just rest and sometimes the creative energy flows late into the night.
>
> *KIMBERLEY*

So one of the big challenges of this Key Phase 3 of Moving Successfully Through Spiritual Emergency is learning to cope with this new acute sensitivity. My lifestyle increasingly resembles, whenever possible, that of a retreat, with no television, radio or newspapers, sometimes not leaving the house, other than to go for walks across the fields or through the woods. For someone like me, who needs a very gentle, quiet environment, writing is an ideal occupation.

ONGOING HEALTH PROBLEMS

This is a major issue in this Key Phase 3. What our bodies and nervous systems went through during spiritual emergency was so extreme, that many of us are left with on-going health problems that stem from that period. Here is Al's account:

> I have never fully recovered the physical or emotional vitality I previously enjoyed and the sensitivity I had towards others' suffering is now much increased. All of this affects the choices I make in my day-to-day life. I have to choose carefully where to invest my available physical and emotional energies. ... It may seem strange to say that I regard my Spiritual Crisis as an ongoing healing experience even though on the face of it, I am less able to cope with ordinary life; but that is how I regard it ... I feel more at peace with myself and the world.

For my part, my immune system took quite a battering from the prolonged stress of my 2006 crisis, leaving me with an ongoing struggle with systemic candida. This is the fungal overgrowth of gut flora that can get into the blood stream, creating fatigue, aching and food intolerances. I also have a lower back problem that stems from then. Kimberley expresses the frustration that you may well resonate with:

> I have great inner strength, determination, vision, creativity and multiple resources but my health has been left somewhat fragile. This continues to be a source of helplessness and frustration to me, particularly as my calling, purpose and creativity are so strong. I get tired very easily, over-stimulated by noise, conversation and time pressures. At best I feel 'OK', but most days I feel like I'm coming out of the tail end of flu.
>
> After my retreat months I felt fantastic, as anyone would. Re-entering 'the world' has steadily weakened me, along with multiple bereavements. I can replenish and restore but it takes much longer these days. I currently have various hormone-related conditions that I am seeking help with.

For some their spiritual crisis itself takes the form of an illness, like ME for example.

Years of being house-bound or even bed-bound can be a process of stripping us down to the bare bones of life, clearing us out and paring down our priorities to what really counts in life. An inner stillness and deep wisdom can be the result.

> I'd been working on my spiritual development for ten years or so and I'd been ill for eight of those when I first met spiritual crisis. ... The strength I've found as a result of having some profound Dark Night of the Soul experiences during my illness has given rise to a deep devotional leading to be of service to others.
>
> *EMILY*

And so we come full circle. Of all the themes to emerge during this Key Phase 3 of Moving Successfully through Spiritual Emergency, we come back to this heart-felt desire to be of service to others. In Hero's Journey terms, this is the test of the true Hero, whether we are ready to share the hard won Elixir with others. Other key themes, such as the need for balance in all areas of our lives, increased sensitivity, the requirement to look after ourselves, whilst also wanting to be of service, these all become central to our well-being.

We have focused in this chapter, for practical purposes, on the key challenges of Going Back Out into the World. The rewards are, however, just as important. Without these we would not feel encouraged to face the challenges, to work with them. We might not want to step back out into the world. This is Kate:

> The depression aspect felt like an enforced shutdown, like a computer that can't be used whilst it's being upgraded. It's like having the addictions we gorge on to fill inner emptiness forcibly removed and being replenished from within instead. Again, paradoxes come into play. On the one hand, it's a cyclical process of ego death and rebirth into expanded psychological and spiritual potentials. On the other, it clarifies, lightens and simplifies. I feel humble and reduced in one sense, expanded in another. Less becomes more.

For many people their lives after spiritual emergency look totally different from before. One friend commented that I had completely re-invented my life. With my previous existence as an academic lecturing in European Studies now a million miles away, I guess I have. I have created a lifestyle that works for me, that nourishes me, that is on my terms. And that must surely be the answer to the tricky question of how to re-engage with the world. It has to be on our terms, not on society's. As Kate says:

At last I am better able to accept the kind of person I am at a time when paradoxically in the eyes of society I am very little. My current circumstances would have catapulted the 20-year old me into depression if she'd had a crystal ball — single, childless and stacking shelves to make ends meet. Yet inside I feel like a success story. My achievements and rites of passage have been inner rather than outer and they feel solid and real.

The process of Going Back Out into the World can take many, many years of major upheavals, adjustments and fine tunings. It can also be interrupted by further descents into the underworld of the dark night of the soul or by further ascents to the heights of holy madness or mystical psychosis. And so the cycle of the Three Key Phases of Moving Successfully Through Spiritual Emergency continues, each time leading to greater richness, greater awakening; each time leading, as Kate so eloquently puts it, to 'expanded psychological and spiritual potentials'.

The Dark Night
of the Globe

We have entered a period of global emergency

ERVIN LASZLO

GLOBAL SPIRITUAL EMERGENCY

This book has been about spiritual emergency and how to move successfully through it at the personal level. And it has been very personal; many have generously shared intimate details in order to help others. Yet this book is also about spiritual crisis at the world level. I mentioned at the outset that part of my aim was to explore the level of intensity we are likely to see in the coming years. Collectively, we have entered global spiritual emergency.

Christopher Bache, who worked for many years with Grof, believes it will be the ecological crisis that drives this spiritual transformation forward. He refers to it as the "melt-down of civilization as we know it".[123] What makes the environmental crisis so lethal, he says, is that it is embedded in the very fabric of our modern existence, every aspect of our lives contributing to it every day. We are on the brink of world breakdown of unprecedented and unimaginable proportions.

How can we not meet such assertions with terror in our hearts, with despair in our minds? When an individual, you or me or a friend, a client, goes through spiritual emergency this is what it is like for us. We have seen this throughout the book. With terror in our hearts and despair in our minds, our world as we know it collapses. Everything falls apart, falls away. We have to be prepared to let go of everything we knew to be familiar, to be of comfort.

What we are facing globally is no different, just on a much larger scale. The microcosm of individual spiritual awakening is reflected in the macrocosm of humanity's awakening of consciousness. It is all part of the same process. As that process inexorably speeds up, as many feel it is, so it tips over into crisis; the crisis of death followed by rebirth, of breaking down in order to break through. What we are beginning to witness is 'the dark night of the globe'. And this is where the lessons of spiritual emergency at the individual level are so valuable. We can know that the apparent breakdown of the 'dark night of the globe' can be followed by breaking through to a whole new level of awakened consciousness. This, for me, is a large part of the value of understanding spiritual crisis at the personal level. We can draw much hope and encouragement from what we learn there.

This sounds like a message of doom and gloom, but again I would remind you of what we have repeatedly seen. People do not regret going through spiritual emergency. The rewards far outweigh the torment, even if it does not feel like it at the time. Humanity will not regret this period of global spiritual emergency.

THE HERO'S JOURNEY OF HUMANITY

Humanity is embarked on its very own Hero's Journey. At this world level too, the model of the Hero's Journey can be very useful. Just as it can help us map the territory individually, it can do the same globally. The Call to Adventure came many years ago with the first warnings of environmental crisis, global warming and peak oil. A large part of humanity Refused the Call. As we have seen, the Call to Adventure often involves painful loss of one kind or another. The Hero was not ready to respond. Although more people seem to be heeding the Call now, we are still very slow, for example, to give up our dependence on our cars. Yet the more the Hero resists the more painful the process is.

As we have also seen, the Call may need to be repeated more loudly or clearly, maybe with further loss, before the Hero is ready to heed it. We have seen an enormous loss of human life with catastrophes like the 2004 tsunami, Hurricane Katrina's devastation of New Orleans and the Japanese tsunami. Slowly we have made the connection between natural disasters, climate change and how we treat our planet. Gradually we are coming to see Gaia as a living being and are changing our relationship to her.

We also know that during personal spiritual crisis 'shadow' material, repressed issues and wounding, can come up from the depths of the psyche to be healed and resolved. The global equivalent of this, the collective shadow, can perhaps be seen manifesting in such phenomena as terrorism.

Following the sequence of the Hero's Journey, the international financial crisis can perhaps be seen as Crossing the First Threshold, the first major challenge or crisis point. During that crisis we came very close to a complete meltdown of our monetary system. Some have taken the opportunity to reassess values and priorities.

One thing we know from individual spiritual emergency, is that as we embark on the journey, the process starts to speed up at an alarming rate. There is no doubt that we are rapidly heading towards the Supreme Ordeal, a global breakdown and breakthrough of unparalleled proportions. The year 2012 has been named by many as the crucial time, but in fact the whole period following that year is a window of opportunity. The Mayan Elders, with their deep wisdom and knowledge of the cycles of the earth and of mankind, have been performing ceremonies to aid the process.

It is important to see the period of chaos, natural disasters and destruction that probably lies ahead as a process of purification, that is needed in order for us to shift to a completely new level of consciousness. Hildegard of Bingen prophetically wrote about "the great tempest, which was purifying the world".[124] We know from experiences of personal spiritual crisis how intense and terrifying this process can be. We also know that bearing with it bears fruit. It is the necessary death before the rebirth. Individually, it is often a symbolic death, an ego death. It may be, however, that unfortunately we see the kind of large-scale deaths and tragedy that we have already seen with the natural disasters of recent years.

If we are prepared for the extreme challenges that lie ahead, we stand a much better chance of being able to move through them successfully, to see them as the wonderful opportunity that they are. We can remember the Chinese character for crisis, made up of the symbols for danger and opportunity.

All those who have been through the nightmare of spiritual emergency and know it to be their greatest blessing, can be a source of encouragement and hope. Knowing that such extreme experiences bring healing, spiritual fulfillment and awakening, can help steady us through the transition of the dark night of the globe.

I leave you with this anonymous poem:

> Love came in the morning.
> Standing as still as light...
> How could I have dreamt of such a dawning
> After so dark a night.

The UK Spiritual Crisis Network

Since 2004, a group of us has been working to raise awareness and understanding about spiritual crisis, to help reduce the suffering caused by it. The Spiritual Crisis Network (SCN) was born directly out of our personal experience of such suffering and the heart-felt desire to help others going through it. How the Network came to be will, however, always remain something of a mystery to me.

Sure, I can tell you about the conference at which the Network was born. I can tell you about our first telephone conference calls, which was how we 'met' initially, because we were scattered all around the country. I can tell you about the two subsequent conferences on spiritual emergency, when more people joined the core group. All that I can tell you.

What I cannot tell you, what I cannot explain, is the wonderful and mysterious way the Universe brought us all together; how the seed of a vision has become something concrete and real. It is not as if there were some great ten-year master plan of how the Network would develop. Yet, here we are, soon, before we know it, celebrating our tenth anniversary. I have found the whole experience of initiating the Spiritual Crisis Network very humbling. It is almost as if it needed to happen, the Universe intended for it to happen. Someone had to be the vehicle for that and I happened to come along, with all the right credentials, including having spent a month on an acute psychiatric ward at the age of 20.

As it turned out, I was not the only one who felt we needed some sort of equivalent in the UK to the American Spiritual Emergence Network. Those who were drawn to forming the core group were 'experiencers' and mental health professionals working with those going through or recovering from spiritual crisis.

From the outset then, the core team has been made up of those with personal and those with professional expertise. Many have both, which is invaluable. This has given the SCN its distinctive quality and strength. Bringing the two together brings a richness and a particular depth and wealth of expertise. Over the years members of the core group have come and gone, but that special mix has remained and is very important to our ethos.

In 2008 we formalized our legal structure, registered as a company and established a Board of Directors. This in turn enabled us, thanks to the hard work of a couple of Directors in particular, to gain charitable status in 2009.

For several years now we have been operating a rota system of responding to emails. Some of these are enquiries, requests for information; many are from people struggling with their experience. Whilst we have to be very realistic about what we can offer, we know from their replies that people are grateful. They feel heard and reassured, despite there being no easy answers. Our volunteers do a fantastic job.

We have several local groups around the country and our vision is to have many more of these. Again, these tend to wax and wane, depending on the time, energy and commitments of those involved. The whole Network is currently organized on a voluntary basis.

SUPPORTING THE SCN

There are two main ways you can support the work of the SCN. One is by becoming a Friend of the Network. This involves making a monthly donation by standing order, which can be as little or as much as you can afford. You can download a form from the website. You can also make donations online.

The other way is by becoming involved with your local group or maybe even setting one up, if there is not one in your area yet. This is a great way of both giving and receiving. Meeting up regularly with others who share such powerfully transformative experiences is very nourishing. You can email info@SpiritualCrisisNetwork.org.uk about this.

RAISING AWARENESS ABOUT SPIRITUAL EMERGENCY

If you have found this book helpful and informative, give a copy to others who need to have an understanding of the issues. That may be family and friends; it may be health professionals, those involved in pastoral care or a whole range of others. Depending on the individual, you might like to recommend one of the academic texts, like the Royal College of Psychiatrist's *Spirituality and Psychiatry* or Isabel Clarke's *Psychosis and Spirituality*. Thank you.

Together we can raise awareness and understanding of spiritual emergency.

www.SpiritualCrisisNetwork.org.uk
www.in-case-of-spiritual-emergency.blogspot.com

Resources

SPIRITUAL EMERGENCE(Y) NETWORKS

www.diabasis.cz
Czech Republic, website in Czech
Email: info@diabasis.cz

www.senev.de
Germany, website in German
Email: info@senev.de

www.SpiritualCrisisNetwork.org.uk
Spiritual Crisis Network, UK
Email: info@SpiritualCrisisNetwork.org.uk

www.spiritualemergence.info/
Spiritual Emergence Network, USA
This website includes an online referral system to mental health professionals who hold a psychospiritual perspective.

www.spiritualemergence.org.au
Spiritual Emergence Network, Australia
Tel: (02) 6624 5037
Email: info@spiritualemergence.org.au

www.spiritualemergence.net
Spiritual Emergence Service (SES), Canada, Facilitated by Janet Taylor
Tel: (604) 687 46 55
Email: ses@spiritualemergence.net

BLOGS

www.in-case-of-spiritual-emergency.blogspot.com
Catherine G. Lucas's blog, exploring themes in this book.

WEBSITES

www.academywisdom.co.uk

The Academy of Living Wisdom website has details of courses run by Catherine G. Lucas, including international telephone seminars.

www.alisterhardytrust.org.uk

Alister Hardy Religious Experience Research Centre & Society, Department of Theology & Religious Studies, University of Lampeter.

www.annabelhollis.co.uk

Annabel, one of the contributors to this book, facilitates groups on the themes of creativity and dreams and is a founding member of the Spiritual Crisis Network.

www.bacp.co.uk

The British Association for Counselling & Psychotherapy

www.bemindful.co.uk

A mindfulness resource, including courses, run by the UK Mental Health Foundation.

www.breathworks-mindfulness.co.uk

Breathworks is an international organization offering Mindfulness courses for individuals as well as training and accreditation for trainers. The website includes a directory of courses available, including those led by Catherine G Lucas.

www.buss.org.uk

British Union of Spiritist Societies

www.conscious.tv

The programme directory/archive includes an interview on spiritual crisis with Fransje De Waard, Frances Goodhall and Annabel Hollis.

www.EmmaBragdon.com

Author of *The Call of Spiritual Emergency*

www.hazelcourteney.com

Hazel Courtney is the author of two books on spiritual emergency

www.iands.org

The International Association for Near-Death Studies

www.kaia.ca

Kaia Nightingale is the author of *Journey Through Transformation* and director of the documentary film *Spiritual Emergency*.

www.kaia.ca/ISER_Home.php

International Spiritual Emergence Resource website 'to help alleviate distress due to profound spiritual transformation and spiritual emergency'.

www.kimberleyjones.com

Kimberley, one of the contributors to this book, 'uses online technologies, intuitive services, art and writing to support and inspire awakening souls all over the world'.

www.kundaliniguide.com

The website is run by Bonnie Greenwell, author of *Energies of Transformation: A Guide to the Kundalini Process*. She offers support for Kundalini awakening on a donation basis.

www.maytree.org.uk

Maytree in London, 'a sanctuary for the suicidal'. Tel: 0207 263 7070

www.mentalhealthrecovery.com

Mary Ellen Copeland's WRAP (Wellness Recovery Action Plan) website

www.psychotherapy.org.uk

United Kingdom Council for Psychotherapy (UKCP)

www.rcpsych.ac.uk/spirit

Spirituality and Psychiatry Special Interest Group, Royal College of Psychiatrists

www.samaritans.org

Tel: 08457 90 90 90

www.scispirit.com/psychosis_spirituality

Discussion group run by Isabel Clarke.

www.sgny.org

The Spiritist Group of New York website includes a directory of Spiritist Centers around the world, including Australia, Canada, Europe and the USA.

www.soulpsychology.ning.com

Transpersonal psychology social network

www.spiritrelease.com

The Spirit Release Foundation, UK

www.spiritreleasement.org

The website of Rev Judith Baldwin, USA

www.spiritualcompetency.com

This website is run by David Lukoff, Co-President of the Association for Transpersonal Psychology, USA.

www.umassmed.edu

The University of Massachusetts Medical School where the Center for Mindfulness, founded by Jon Kabat-Zinn is based.

CDS AND DVDS

Evolving Minds DVD
> An exploration of the alternatives to psychiatry and the link between psychosis and spirituality. Directed by Mel Gunasena and Produced by Undercurrents
> hello@undercurrents.org
> www.undercurrents.org/minds

Psychospiritual Crisis: Where Mysticism and Mental Health Meet Audio MP3
> Recording of a talk by Catherine G Lucas from a conference at Norfolk & Waveney Mental Health Care Partnership NHS Trust (UK).
> Available at www.in-case-of-spiritual-emergency.blogspot.com

Spiritual Emergency DVD
> Documentary film by Kaia Nightingale
> Available from www.kaia.ca

Understanding Spiritual and Mental Health Audio CD
> Seminar given by Divaldo Franco on Spiritism and Mental Health.
> Available from Amazon

Body Scan CD
> There are many different versions of the Body Scan, a relaxing body awareness practice. One is available from *www.breathworks-mindfulness.co.uk*

PLACES TO STAY

The following places have been selected as suitable either because they specifically state that they support those going through or recovering from spiritual emergency or because they are known personally to members of the Spiritual Crisis Network. Despite the individual spiritual affiliation of each center, all are open to any faith. It is important that you do your own research, speak to staff and get a sense of where feels right for you.

A. Places Suitable for those in Crisis, offering 24-hr support

MAYTREE — a sanctuary for the suicidal

> **Referral:** self-referral or family
> **Who they take:** those who are actively suicidal
> **How long you can stay:** maximum of 4 nights
> **Type of support offered:** 24-hour listening and support

Accommodation/food offered: single room, three meals a day
Cost: free
Address: North London
Contact details: tel 020 7263 7070, email: maytree@maytree.org.uk
Spiritual affiliation: none
Website: www.maytree.org.uk

THE RETREAT

Referral: Doctor/health professional
Who they take: those experiencing 'periods of mental ill health'
How long you can stay: long term
Type of support offered: inpatient and outpatient / a range of psychotherapies, individual or group / medical therapies / psychosocial education
Accommodation/food offered: single rooms, three meals a day
Cost: approx. £400 a day
Address: Heslington Road, York YO10 5BN
Contact details: tel 01904 412551 ext.2304, email: info@theretreatyork.org.uk
Spiritual affiliation: 'respect for the spiritual needs of the individual' (Quaker founded)
Website: www.retreat-hospital.org
Other notes: contact *The Retreat* for details of possible funding

CHERRY ORCHARDS COMMUNITY

Referral: social worker
Who they take: those suffering mentally who feel they can benefit from what is offered
How long you can stay: up to two years
Type of support offered: 24-hour
Accommodation/food offered: single room/3 meals a day
Cost: negotiated with placing authority, social services or health trust
Address: Canford Lane, Westbury-on-Trym, Bristol
Contact details: tel 0117 950 3183
Spiritual affiliation: anthroposophical/Rudolf Steiner approach
Website: www.cherryorchards.co.uk
Other notes: this is a therapeutic community aiding individual transition from crisis to development

B. Places Suitable for those in post-crisis,
offering some support

CENTRESPACE

Referral: self-referral, family, Doctor, clergy, mental health team
Who they take: those in need of quiet time and space
How long you can stay: short term or long term
Type of support offered: listening/counsellingavailable
Accommodation/food offered: either self-catering or meals provided
Cost: dependent on number of meals, £35-40 per night full board,
Address: Canterbury, Kent
Contact details: tel 01227 462038, email: podger@centrespace.freeserve.co.uk
Spiritual affiliation: Christian and Interfaith

LOTHLORIEN THERAPEUTIC COMMUNITY

Referral: self-referral or through social worker
Who they take: 'people with mental health problems'
How long you can stay: up to 2 years
Type of support offered: individual and group staff on call evenings and week-ends
Accommodation/food offered: 3 vegetarian meals a day
Cost: £405 per week
Address: Corsock, Castle Douglas, Dumfries and Galloway DG7 3DR, Scotland
Contact details: tel 01644 440 602, email: lothlorien1@btopenworld.com
Spiritual affiliation: Buddhist
Website: www.lothlorien.tc/

SACRED SPACE FOUNDATION

Referral: self-referral
Who they take: those wanting to retreat with support and guidance
How long you can stay: negotiable
Type of support offered: personal and spiritual counselling
Accommodation/food offered: mainly self-catering
Cost: there are no fixed charges, donations invited
Address: Penrith, Cumbria

Contact details: tel 017684 86868, email: jeannie@sacredspace.org.uk
Spiritual affiliation: interfaith
Website: www.sacredspace.org.uk

ACADEMIC RESEARCH & JOURNAL ARTICLES

Brett, Caroline. "Transformative Crises", in Isabel Clarke (ed), *Psychosis and Spirituality: Consolidating the New Paradigm.* Chichester: Wiley-Blackwell, 2010. Pp. 155-174.

Clarke, Isabel (ed.). "Special Issue: Taking Spirituality Seriously", *The Journal of Critical Psychology, Counselling & Psychotherapy*, vol. 2, no. 4 (winter 2002).

Collins, Mick. "Spiritual Emergency & Occupational Identity: a Transpersonal Perspective", *British Journal of Occupational Therapy,* vol. 70, no. 12 (Dec 2007). Pp. 504-512.

Greyson, Bruce. "The Near-death Experience Scale: Construction, Reliability, and Validity", *Journal of Nervous and Mental Disease*, vol. 171, no. 6 (June 1998). Pp. 369-375.

Greyson, Bruce. "The Physio-Kundalini Syndrome and Mental Illness", *Journal of Transpersonal Psychology*, vol. 25, no. 1 (1993). Pp. 43-58.

Lukoff, David. "From Spiritual Emergency to Spiritual Problem: the transpersonal roots of the new DSM-IV category", *Journal of Humanistic Psychology,* vol. 38, no. 2 (1998). Pp. 21-50. Also available online at www.spiritualcomptency.com.

Read, Tim and Nicki Crowley. "The Transpersonal Perspective", in Chris Cook, Andrew Powell, Andrew Sims (eds.), *Spirituality and Psychiatry.* London: RC Psych Publications, 2009. Pp. 212-232.

Stevenson, Ian. "Phobias in Children Who Claim to Remember Previous Lives", *Journal of Scientific Exploration,* vol. 4 (1990). Pp. 243-254.

Stevenson, Ian. "Some of My Journeys in Medicine", *The Flora Levy Lecture in the Humanities,* University of Southwestern Louisiana (1989).
Also available online at www.medicine.virginia.edu under some-of-my-journeys-in-medicine.pdf.

Further Reading

SPIRITUAL EMERGENCY

Assagioli, Roberto. *Psychosynthesis*. Amherst: The Synthesis Center, 2000.

Bragdon, Emma. *The Call of Spiritual Emergency*. San Francisco: Harper & Row, 1990.

Courteney, Hazel. *Divine Intervention*. London: CICO Books, 2002.

Elam, Jennifer. *Dancing with God Through the Storm: Mysticism and Mental Illness*. Media: Way Opens Press, 2002.

Grof, Stanislav & Christina (eds.). *Spiritual Emergency: When Personal Transformation Becomes a Crisis*. New York: Penguin Putnam, 1989.

——. *The Stormy Search for the Self*. New York: Penguin Putnam, 1990.

Jung, Carl. *The Red Book: Liber Novus*. Sonu Shamdasani (ed). New York: W. W. Norton & Co, 2009.

May, Gerald G. *The Dark Night of the Soul*. New York: HarperOne, 2005.

Nightingale, Kaia. *Journey Through Transformation*. Ottawa: Baico, 2007.

Perry, John Weir. *The Far Side of Madness*. Putnam: Spring Publications, 2005.

——. *Trials of the Visionary Mind*. Albany: State University of New York Press, 1999.

St. John of the Cross. *The Dark Night of the Soul*. Mineola: Dover Publications, 2003.

de Waard, Fransje. *Spiritual Crisis: Varieties and Perspectives of a Transpersonal Phenomenon*. Exeter: Imprint Academic, 2010.

BIOGRAPHY

Swami Amritasvarupananda. *Mata Amritanandamayi: A Biography*. Mata Amritanandamayi, Kollam: Mission Trust, 1988.

St. Teresa of Ávila. *The Life of St. Teresa of Ávila by Herself*. Trans. J. M. Cohen. London: Penguin Books, 1957.

Changchub, Gyalwa and Namkhai Nyingpo. *Lady of the Lotus-Born: The Life and Enlightenment of Yeshe Tsogyal*. Boston: Shambhala, 2002.

Cornell, Judith. *Amma: A Living Saint*. London: Piatkus, 2001.

Degler, Teri. *St. Hildegard of Bingen*. Markdale: Institute of Consciousness Research, 2007.

Dunne, Claire. *Carl Jung: Wounded Healer of the Soul*. New York: Parabola Books, 2000.

Edwards, Cliff. *Van Gogh and God*. Chicago: Loyola University Press, 1989.

van Gogh, Vincent. *The Complete Letters of Vincent van Gogh*. Volume II, 2nd edition. London: Thames & Hudson, 2000.

Jung, Carl Gustav. *Memories, Dreams, Reflections*. London: Fontana Paperbacks, 1983.

Keenan, Brian. *An Evil Cradling*. London: Vintage, 1993.

Maugham, Somerset. "The Saint" in *Points of View*. London: Vintage, 2000.

Natarajan, A R. *Timeless in Time: Sri Ramana Maharshi*. Bangalore: Ramana Maharshi Centre for Learning, 1999.

Powers Erickson, Kathleen. *At Eternity's Gate: The Spiritual Vision of Vincent van Gogh*. Grand Rapids: Eerdmans Publishing Co, 1998.

Somé, Malidoma Patrice. *Of Water and the Spirit*. New York: Arkana, 1995.

Sundaram, Pingali Surya (ed.). *Sri Ramana Leela*. Tiruvannamalai: Sri Ramanasramam, 2004.

MINDFULNESS

Analayo. *Satipatthana: The Direct Path to Realization*. Birmingham: Windhorse Publications, 2003.

Burch, Vidyamala. *Living Well with Pain and Illness*. London: Piatkus, 2008.

Kabat-Zinn, Jon. *Full Catastrophe Living*. New York: Dell Publishing, 1990.

Nhat Hanh, Thich. *The Miracle of Mindfulness*. London: Rider, 1991.

Williams, Mark, and John Teasdale, Zindel Segal, Jon Kabat-Zinn. *The Mindful Way Through Depression*. New York: The Guilford Press, 2007.

OTHER

Atherton, Mark. *Hildegard of Bingen: Selected Writings*. London: Penguin Books, 2001.

Bache, Christopher M. *Dark Night, Early Dawn*. Albany: State University of New York Press, 2000.

Bailey, Lee W and Jenny Yates. *The Near-Death Experience: A Reader*. New York: Routledge, 1996.

Bingen, Hildegard of. *Scivias*. Trans. Mother Columba Hart and Jane Bishop. Mahwah: Paulist Press, 1990.

Bowie, Fiona and Oliver Davies (eds.). *Hildegard of Bingen: An Anthology*. London: SPCK, 1990.

Bragdon, Emma. *Kardec's Spiritism*. Woodstock: Lightening Up Press, 2004.

Bucke, Richard Maurice. *Cosmic Consciousness.* Philadelphia: Innes and Sons, 1901.

Cameron, Julia. *The Artist's Way.* London: Souvenir Press, 1994.

Campbell, Joseph. *The Hero with a Thousand Faces.* London: Fontana Press, 1993.

Clarke, Isabel (ed.). *Psychosis and Spirituality: Consolidating the new paradigm.* Chichester: Wiley-Blackwell, 2010.

Cook, Chris, and Andrew Powell, Andrew Sims (eds.). *Spirituality and Psychiatry.* London: RCPsych Publications, 2009.

Fenwick, Peter and Elizabeth. *The Truth in the Light: Investigation of over 300 Near-Death Experiences.* London: Headline, 1995.

James, William. *The Varieties of Religious Experience.* New York: Penguin, 1982.

Jaspers, Karl. *Strindberg and Van Gogh.* Trans. Oskar Grunow and David Woloshin. Tucson: University of Arizona Press, 1977.

Khalsa, Gurmukh Kaur, and Andrew Newberg, Sivananda Rhada, Ken Wilber, John Selby (eds). *Kundalini Rising: Exploring the Energy of Awakening.* Boulder: Sounds True, 2009.

Kornfield, Jack. *After the Ecstasy, the Laundry.* New York: Bantom Books, 2000.

Levine, Peter. *Waking The Tiger: Healing Trauma.* Berkeley: North Atlantic Books, 1997.

Linn, Denise. *Signposts: How to interpret the coincidences and symbols in your life.* London: Rider, 1996.

Mijares, Sharon (ed.). *Psychospiritual clinician's handbook: Alternative methods for understanding and healing mental disorders.* Binghamton: Haworth Press, 2005.

Miller, Joe and Guinevere (eds.). *The Spiritual Teaching of Ramana Maharshi.* Boulder: Shambhala, 1972.

Monroe, Robert. *Far Journeys.* Garden City: Doubleday, 1985.

——. *Journeys Out of the Body.* New York: Broadway Books, 2001.

——. *The Ultimate Journey.* Garden City: Doubleday, 1994.

Nelson, John E. *Healing the Split.* Albany: State University of New York Press, 1994.

Newton, Michael. *Journey of Souls: Case Studies of Life Between Lives.* St. Paul: Llewellyn, 1994.

——. *Destiny of Souls: New Case Studies of Life Between Lives.* St. Paul: Llewellyn, 2000.

Podvoll, Edward M. *Recovering Sanity: a compassionate approach to understanding and treating psychosis.* Boston: Shambhala Publications, 2003.

Tolle, Eckhart. *A New Earth.* London: Penguin Books, 2005.

——. *The Power of Now.* London: Hodder & Stoughton, 2001.

Tomlinson, Andy. *Exploring the Eternal Soul.* Winchester: O Books, 2007.

Tucker, Jim. *Life Before Life.* London: Piatkus, 2005.

Vogler, Christopher. *The Writer's Journey.* London: Boxtree Ltd, 1996.

Woolger, Roger J. *Other Lives, Other Selves.* London: Thorsons, 1999.

Footnotes & References

1   From a paper presented by Prof David Fontana at *Mental Health, Wellbeing and Spirituality* conference, NHS Tayside, Dundee, August 2004.

2   Roberto Assagioli, *Psychosynthesis,* The Synthesis Centre, 2000, pp. 39-41.

3   From Margot Anand, *The New Art of Sexual Ecstasy,* Thorsons, 2003.

4   Jack Kornfield, "Obstacles and Vicissitudes in Spiritual Practice", in Stanislav & Christina Grof (eds.), *Spiritual Emergency: When Personal Transformation Becomes a Crisis,* Penguin Putnam, 1989, p. 159.

5   Emma Bragdon, *The Call of Spiritual Emergency,* Harper and Row, 1990, pp. 85-86.

6   Reported in Kaia Nightingale, *Journey Through Transformation,* Baico, 2007.

7   For example, Dean Radin, *The Conscious Universe: the Scientific Truth of Psychic Phenomena,* HarperCollins, 1997.

8   For example, Charles Tart (ed.), *Body Mind Spirit: Exploring the Parasychology of Spirituality,* Hampton Roads Publishing Company, 1977.

9   Richard Bucke, *Cosmic Consciousness,* Innes and Sons, 1905, p. 2.

10 Richard Bucke, *Cosmic Consciousness,* p. 3.

11 William James, *The Varieties of Religious Experience,* Penguin, 1982, p. 380.

12 Roberto Assagioli, *Psychosynthesis,* pp. 34-35.

13 Carl Jung, *The Red Book: Liber Novus* Sonu Shamdasani (ed.), W W Norton & Co, 2009, p. 201.

14 Carl Jung, *Memories, Dreams, Reflections,* Fontana Paperbacks, 1983.

15 John Weir Perry, *Trials of the Visionary Mind,* State University of New York Press, 1999, p. 27.

16 *The Laughing Man,* magazine, vol. 5, no. 2, 1984. Interview published under the title "Sparks of Light", transcript available at www.dharmacafe.com and also held in University of Glasgow Library Special Collections.

17 R.D. Laing, "Transcendental Experience in Relation to Religion and Psychosis", in Stanislav & Christina Grof (eds.), *Spiritual Emergency,* Penguin Putnam, 1989, p. 60.

18 Perry, *Trials of the Visionary Mind,* p. vii.

19 Perry, *Trials of the Visionary Mind,* p. 20.

20 Assagioli, *Psychosynthesis,* p. 56.

21 John Weir Perry, *The Far Side of Madness,* Spring Publications, 2005, p. 173.

22 Perry, *Trials of the Visionary Mind,* p. 164.

23 Gurmukh Kaur Khalsa, et al. (eds.), *Kundalini Rising: Exploring the Energy of Awakening,* Sounds True, 2009, p. 101.

24   Khalsa, et al. (eds.), *Kundalini Rising*, pp. 175-176.

25   Gopi Krishna, *Kundalini: The Evolutionary Energy in Man*, Shambhala, 1971.

26   See, for example, Lee Sannella, *The Kundalini Experience*, Integral Publishing, 1987 or Lee Sannella, "Kundalini: Classical and Clinical", in Stanislav & Christina Grof (eds.), *Spiritual Emergency: When Personal Transformation Becomes a Crisis*, Penguin Putnam, 1989.

27   Itzhak Bentov, *Stalking the Wild Pendulum: On the Mechanics of Consciousness*, Inner Traditions, 1977.

28   Bonnie Greenwell, *Energies of Transformation: A Guide to the Kundalini Process*, Shakti River Press, 1995.

29   Khalsa, et al. (eds.), *Kundalini Rising: Exploring the Energy of Awakening*.

30   Bruce Greyson, "Near-Death Experiences and the Physio-Kundalini Syndrome", in Gurmukh Kaur Khalsa, et al. (eds.), *Kundalini Rising: Exploring the Energy of Awakening*, Sounds True, 2009.

31   Gurmukh Kaur Khalsa, et al. (eds.), *Kundalini Rising*, p. 127.

32   See, for example, P. M. H. Atwater, *Beyond the Light*, Avon Books, 1995, or Betty Eadie, *Embraced By the Light*, Bantam, 1994.

33   Robert Monroe, *Journeys Out of the Body*, Anchor Press, 1970.
Robert Monroe, *Far Journeys*, Doubleday, 1985.
Robert Monroe, *The Ultimate Journey*, Doubleday, 1994.

34   See, for example, Elisabeth Kübler-Ross, *On Life After Death*, Celestial Arts, 1991.

35   See, for example, Raymond Moody, *Life After Life*, HarperSanFrancisco, 2001 (2nd ed.).

36   See, for example, Janice Miner Holden, Bruce Greyson, Debbie James (eds.), *The Handbook of Near-death Experiences: Thirty Years of Investigation*, ABC-CLIO, 2009.

37   See, for example, Peter and Elizabeth Fenwick, *The Truth in the Light: Investigation of over 300 Near-Death Experiences*, Headline Book Publishing, 1995.

38   See Horizon Research Foundation, www.horizonresearch.org.

39   Lee Bailey and Jenny Yates (eds.), *The Near-Death Experience: A Reader* Routledge, 1996.

40   Emma Bragdon, *Kardec's Spiritism*, Lightening Up Press, 2004, p. 146.

41   Peter Levine, *Waking the Tiger: Healing Trauma*, North Atlantic Books, 1997.

42   See, for example, Michael Newton, *Journey of Souls: Case Studies of Life Between Lives*, Llewellyn, 1994. And Michael Newton, *Destiny of Souls: New Case Studies of Life Between Lives*, Llewellyn, 2000.

43   Andy Tomlinson, *Exploring the Eternal Soul*, O Books, 2007.

44   See, for example, Ian Stevenson, *Where Reincarnation and Biology Intersect*, Praeger Publishers, 1997.

45   Jim Tucker, *Life Before Life*, Piatkus, 2005.

46   See the Spirit Release Foundation, www.spiritrelease.com.

47   Bragdon, *Kardec's Spiritism*.

48   For details see Second British Congress on Medicine and Spirituality, (2009), www.buss.org.uk.

49   John E. Nelson, *Healing the Split*, State University of New York Press, 1994.

50 Chris Cook, Andrew Powell, Andrew Sims (eds.), *Spirituality and Psychiatry*, The RCPsych Publications, 2009.

51 Diagnostic and Statistical Manual of Mental Disorders, 4th edition, American Psychiatric Association, 1994, p. 685.

52 See David Lukoff, "Kundalini Awakening" in Gurmukh Kaur Khalsa, et al. (eds.), *Kundalini Rising: Exploring the Energy of Awakening*, Sounds True, 2009.

53 Assagioli, *Psychosynthesis*, p. 51.

54 Assagioli, *Psychosynthesis*, p. 51.

55 Nelson, *Healing the Split*.

56 Jon Kabat-Zinn, *Full Catastrophe Living*, Dell Publishing, 1990.

57 Mark Williams, John Teasdale, Zindel Segal and Jon Kabat-Zinn, *The Mindful Way Through Depression*, The Guilford Press, 2007.

58 Paul Chadwick, *Person-Based Cognitive Therapy for Distressing Psychosis*, John Wiley & Sons, 2006.

59 Nelson, *Healing the Split*.

60 Fiona Harrold, *Be Your Own Life Coach*, Hodder and Stoughton, 2001.

61 Gyalwa Changchub and Namkhai Nyingpo, *Lady of the Lotus-Born: The Life and Enlightenment of Yeshe Tsogyal*, Shambhala, 2002, p. 73.

62 Changchub and Nyingpo, *Lady of the Lotus-Born*, p. 74.

63 Changchub and Nyingpo, *Lady of the Lotus-Born*, p. 75.

64 Changchub and Nyingpo, *Lady of the Lotus-Born*, p. 143.

65 Changchub and Nyingpo, *Lady of the Lotus-Born*, p. 146.

66 St. Teresa of Ávila, *The Life of St. Teresa of Ávila by Herself*, trans. J. M. Cohen, Penguin Books, 1957, p. 24.

67 St. Teresa of Ávila, *The Life of St. Teresa of Ávila by Herself*, p. 35.

68 Jack Kornfield, "Obstacles and Vicissitudes in Spiritual Practice", in Stanislav & Christina Grof (eds.), *Spiritual Emergency: When Personal Transformation Becomes a Crisis*.

69 St. Teresa of Ávila, *The Life of St. Teresa of Ávila by Herself*, p. 35.

70 St. Teresa of Ávila, *The Life of St. Teresa of Ávila by Herself*, p. 35.

71 St. Teresa of Ávila, *The Life of St. Teresa of Ávila by Herself*, p. 44.

72 St. Teresa of Ávila, *The Life of St. Teresa of Ávila by Herself*, p. 137.

73 St. Teresa of Ávila, *The Life of St. Teresa of Ávila by Herself*, p. 144.

74 Roberto Assagioli, *Psychosynthesis*, p. 43. Quoted in Roberto Ass etc.

75 St. John of the Cross, *The Dark Night of the Soul*, Dover Publications, 2003.

76 Somerset Maugham, The Saint' in *Points of View*, Vintage, 2000, p. 58.

77 Pingali Surya Sundaram (ed.), *Sri Ramana Leela*, Sri Ramanasramam, 2004, p. 22.

78 A. R. Natarajan, *Timeless in Time: Sri Ramana Maharshi*, Ramana Maharshi Centre for Learning, 1999, p. 75.

79 Joe and Guinevere Miller (eds.), *The Spiritual Teaching of Ramana Maharshi*, Shambhala, 1972, Foreword by C G Jung, p. X.

80 Mark Atherton, *Hildegard of Bingen: Selected Writings*, Penguin Books, 2001, pp. 171-172.

81 Atherton, *Hildegard of Bingen*, p. 190.

82 Fiona Bowie and Oliver Davies (eds.), *Hildegard of Bingen: An Anthology*, SPCK, 1990, p. 91.

83   Bowie and Davies (eds.), *Hildegard of Bingen*, p. 68.

84   Bowie and Davies (eds.), *Hildegard of Bingen*, p. 85.

85   Teri Degler, *St Hildegard of Bingen*, Institute of Consciousness Research, 2007.

86   Bowie and Davies (eds.), *Hildegard of Bingen*, pp. 90-91.

87   Atherton, *Hildegard of Bingen*, p. 105.

88   Hildegard of Bingen, *Scivias*, (trans. Mother Columba Hart and Jane Bishop) Paulist Press, 1990, pp. 515-519.

89   Vincent van Gogh, *The Complete Letters of Vincent van Gogh*, Volume II, 2nd edition, Letter 242. Thames & Hudson, 2000.

90   Vincent van Gogh, *The Complete Letters of Vincent van Gogh*, Letter 339a.

91   Karl Jaspers, *Strindberg and Van Gogh*, (trans. Oskar Grunow and David Woloshin) University of Arizona Press, 1977, p. 164.

92   Vincent van Gogh, *The Complete Letters of Vincent van Gogh*, Letter 570.

93   Vincent van Gogh, *The Complete Letters of Vincent van Gogh*, Letter 605.

94   Jaspers, *Strindberg and Van Gogh*, p. 185.

95   Jaspers, *Strindberg and Van Gogh*, p. 186.

96   Vincent van Gogh, *The Complete Letters of Vincent van Gogh*, Letter 604.

97   *C. G. Jung Letters Volume I*, Princeton University Press 1972, p. 377, letter to P. W. Martin, quoted in Claire Dunne *Carl Jung: Wounded Healer of the Soul*, Parabola Books, 2000, p. 3.

98   Carl Jung, *The Red Book: Liber Novus*, Sonu Shamdasani (ed.), W W Norton & Co, 2009.

99   Carl Gustav Jung, *Memories, Dreams, Reflections*, Fontana Paperbacks, 1983, p. 358.

100   Jung, *Memories, Dreams, Reflections*, p. 218.

101   Jung, *Memories, Dreams, Reflections*, p. 214.

102   Jung, *The Red Book*, p. 236.

103   Jung, *The Red Book*, p. 295.

104   Jung, *The Red Book*, p. 241.

105   Jung, *The Red Book*, p. 235.

106   Jung, *The Red Book*, p. 298.

107   Jung, *The Red Book*, p. 295.

108   Jung, *The Red Book*, p. 240.

109   Jung, *The Red Book*, p. 250.

110   Eckhart Tolle, *The Power of Now*, New World Library, 1999, p. 1.

111   From an interview by the publisher Sounds True, www.soundstrue.com

112   Tolle, *The Power of Now*, p. 3.

113   Tyting Community Project website, July 2009 www.tytingcommunityproject.org.uk

114   Swami Amritasvarupananda, *Mata Amritanandamayi: A Biography*, Mata Amrita-nandamayi, Mission Trust, 1988, p. 83 and p. 90.

115   Swami Amritasvarupananda, *Mata Amritanandamayi: A Biography*, p.90.

116   Vidyamala Burch, *Living Well with Pain and Illness*, Piatkus, 2008.

117   Brian Keenan, *An Evil Cradling*, Vintage, 1993, p.78.

118   Jack Kornfield, *After the Ecstasy, the Laundry*, Bantom Books, 2000.

119   Jack Kornfield, "Obstacles and Vicissitudes in Spiritual Practice", in Stanislav & Christina Grof (eds.), *Spiritual Emergency: When Personal Transformation Becomes a Crisis*, p. 160.

120   Denise Linn, *Signposts: How to interpret the coincidences and symbols in your life,* Rider, 1996, p. 116.

121   T. S. Eliot, *Collected Poems 1909-1962,* Faber and Faber, 1974.

122   *Kriyas* are spontaneous body movements that may occur to clear blockages held in the energy body. Most *kriyas* are experienced as simple tremors or twitches that release tension in the body. They can, however, be many and varied and much more complex than simple twitching, including the body moving involuntarily into yoga positions.

123   Christopher Bache, *Dark Night, Early Dawn,* State University of New York Press, 2000, p. 233.

124   Hildegard of Bingen, *Scivias,* p. 515.

Reprinted by kind permission

The author has made every reasonable attempt to locate and contact copyright owners and is grateful for the following:

Dark Night of the Soul by St John of the Cross, translated by E. Allison Peers. Copyright © 2003, Dover Publications. Reprinted by kind permission of Dover Publications.

The Far Side of Madness by John Weir Perry. Copyright © 2005, Spring Publications, Inc. Reprinted by kind permission of Spring Publications, Inc.

Kundalini Rising: Exploring the Energy of Awakening edited by Gurmukh Kaur Khalsa, Andrew Newberg, Sivananda Rhada, Ken Wilber, John Selby. Copyright © 2009, Sounds True. Reprinted by permission of Sounds True, www.soundstrue.com.

Hildegard of Bingen: An Anthology edited by Fiona Bowie and Oliver Davies. Copyright © 1990, SPCK. Reprinted by kind permission of SPCK.

Hildegard of Bingen: Scivias. Copyright © 1990 by The Abbey of Regina Laudis: Benedictine Regina Laudis of the Strict Observance, Inc., Paulist Press, Inc., Mahwah, NJ. Reprinted by kind permission of Paulist Press, Inc., www.paulistpress.com.

The Life of Saint Teresa of Ávila by Herself translated with an introduction by J. M. Cohen (Penguin Classics, 1957). This translation copyright © J. M. Cohen, 1957. Reprinted by kind permission of Penguin Group (UK).

Psychosynthesis by Roberto Assagioli. Copyright © 2000, The Synthesis Center. Reprinted by kind permission of The Synthesis Center.

The Red Book by C.G. Jung, edited by Sonu Shamdasani, translated by Mark Kyburz, John Peck, and Sonu Shamdasani. Copyright © 2009 by the Foundation of the Works of C.G. Jung. Copyright translation © 2009 by Mark Kyburz, John Peck, and Sonu Shamdasani. Used by permission of W.W. Norton & Company, Inc.

Trials of the Visionary Mind: Spiritual Emergency and the Renewal Process by John Weir Perry. Copyright © 1999, State University of New York Press. Reprinted by permission of SUNY Press.

'The Unbroken' by Rashani Réa, 1991. Published in her book, *Beyond Brokenness*. Reproduced by kind permission, www.rashani.com.

Vincent Van Gogh - The Letters edited by Leo Jansen, Hans Luijten and Nienke Bakker. Copyright © 2009 Van Gogh Museum, Amsterdam, Huygens Institute, The Hague, Mercatorfonds, Brussels, and Thames & Hudson Ltd, London. Reprinted by kind permission of Thames & Hudson.

FINDHORN PRESS

Life Changing Books

For a complete catalogue,
please contact:

Findhorn Press Ltd
117-121 High Street,
Forres IV36 1AB,
Scotland, UK

t +44 (0)1309 690582
f +44 (0)131 777 2711
e info@findhornpress.com

or consult our catalogue online
(with secure order facility) on
www.findhornpress.com

For information on the Findhorn Foundation:
www.findhorn.org